THE LIVES TO COME

The Genetic Revolution and Human Possibilities

PHILIP KITCHER

A TOUCHSTONE BOOK

Published by Simon & Schuster

TOUCHSTONE
Rockefeller Center
1230 Avenue of the Americas
New York, NY 10020

First Touchstone Edition 1997

TOUCHSTONE and colophon are registered trademarks
of Simon & Schuster Inc.

Designed by Jeanette Olender

Manufactured in the United States of America

1   3   5   7   9   10   8   6   4   2

Library of Congress Cataloging-in-Publication Data
Kitcher, Philip, date.
The lives to come : the genetic revolution
and human possibilities / Philip Kitcher. —1st Touchstone ed.
p. cm.
"A Touchstone book."
Includes bibliographical references and index.
1. Human genetics—Popular works.
2. Human Genome Project—Popular works.
3. Genetic engineering—Moral and ethical aspects.
I. Title.
QH431.K54   1997
599.93'5—dc21   97-17314
CIP

ISBN 0-684-80055-1
0-684-82705-0 (Pbk)

For Alice, her kindness and her courage,

and in memory of George, his wisdom and his integrity.

Contents

List of Illustrations 9

1. The Shapes of Suffering 13

2. Our Mortal Coils 23

3. To Test or Not to Test? 65

4. The Road to Health? 87

5. A Patchwork of Therapies 105

6. The New Pariahs? 127

7. Studies in Scarlet 157

Interlude: The Specters That Won't Go Away 181

8. Inescapable Eugenics 187

9. Delimiting Disease 205

10. Playing God? 221

11. Fascinating Genetalk 239

12. Self-Dissection 271

13. The Quality of Lives 285

14. An Unequal Inheritance 309

Postscript (March 1997) 327

Notes 343

Glossary 363

Acknowledgments 369

Index 375

List of Illustrations

Figure 2.1.   The Structure of DNA   30

Figure 2.2.   DNA Replication   32

Figure 2.3.   Transcription and Translation   35

Figure 2.4.   Various Kinds of Mutation   37

Figures 2.5a   Schematic Structure of a Chromosome
and 2.5b.   and of a Eukaryotic Gene   42

Figure 2.6.   Restriction Enzymes Cut DNA   45

Figure 2.7.   Mapping Loci   48

Figure 2.8.   RFLPs   51

Figures 2.9a   DNA Sequencing   58–59
and 2.9b.

Figure 6.1.   Different Workplace Conditions   149

Figure 7.1.   The Polymerase Chain Reaction (PCR)   160

Every night & every Morn

Some to Misery are Born.

Every Morn & every Night

Some are Born to sweet delight.

—WILLIAM BLAKE
Auguries of Innocence

1

The Shapes of Suffering

Each of the rooms at Children's Convalescent Hospital in San Diego is alive with color. Around the three (or four, or five) beds, vivid pictures and brightly painted walls provide an exuberant welcome. Large stuffed toys in red, green, or purple sometimes lie scattered on the floor; more frequently they sit stiffly, untouched, the product of adult rearrangement. On the beds lie the children—the two-year-olds, the four-year-olds, the ten-year-olds, and the teenagers—some whose limbs convulse erratically, others who are unnaturally still.

Beside the doors there are often notes, handwritten by the nursing staff, providing detailed instructions on what a particular child has achieved, what can be done to foster further development. One boy is starting to pull himself up if a pair of hands is held out to him in a particular way; a girl smiled when a music box played a certain tune. Records of small accomplishments, they are built on daily, read and revised by a team of nurses whose care permits truncated human lives still to grow, even if, for most observers, there is no pace perceived.

Some of these children will die very young in the bright world of Children's Hospital. For others there will be a succession of hospital rooms, without toys or murals, and a succession of nurses, who may no longer wonder how to extend their hands to elicit a response, who no longer wind music boxes, a succession of rooms, to which the children will be largely oblivious. A few will return, at least for a time, to their parents.

In one of the rooms at Children's Hospital, a four-year-old boy who looks much younger than his age sits on the floor playing with a toy. He is afflicted with neurofibromatosis, and his development was quite normal until a year or so ago, when he suffered a massive seizure, which returned him to a state of early infancy. Slowly, the doctors and nurses at Children's have brought him back, watching him reacquire the capacities of a young toddler. Soon his parents will follow his further retracing of early developmental steps, waiting for the next seizure, wondering how far he will go this time and how much will be lost.

For most of the other children, there are not even vestiges of hope. A doctor judges that one child is left with only a brain stem, the result of being swung from the shoulder like a sack, his head brutally smashed against some hard surface. The parents have not yet signed the release allowing support to cease and the child to die. Another, a teenage boy, rescued from drowning at a sixth-birthday party, has survived without special medical support, quiescent for the better part of a decade. Equally still is a victim of a genetically based neuro-degenerative disease—Canavan's disease—whose blank eyes and wasted legs sum up the poignancy of his eight years of life.

Accidents, complications of pregnancy, birth trauma, violent abuse —and genes, sometimes genes that are known, often genes that are only suspected because the affliction "runs in the family"—stand behind the individual tragedies of Children's. Half a century ago the hereditary diseases would have seemed the truly recalcitrant cases, the ones that must inevitably resist the most thorough efforts at prevention. Today, thanks to the enormous successes of molecular genetics, that judgment has been reversed: We anticipate that deeper understanding of human genes will make it possible to cure genetic diseases or, where that is impossible, to test in advance and to avoid bringing into the world children who will inevitably suffer.

We are in the middle of a scientific revolution, a transformation of our ideas about nature whose only equivalents are the birth of modern science in the seventeenth century and the upheavals in physics of the early twentieth century. Fifty years ago the chemical identity of the hereditary material was unknown. Since then molecular biology has blossomed, fathoming the structure of *DNA (deoxyribonucleic acid)*, cracking the genetic code, cutting, joining, and moving genes, reading the sequences of bases in DNA molecules. Today what was a challenge for the best postdoctoral researchers little more than a decade ago has metamorphosed into everyday work for undergraduates, technicians, and even robots. Sporadic announcements of genetic discoveries have now become commonplace, as newspapers and popular magazines celebrate not the isolated achievement of a year but the "gene of the week," even the "gene of the day."

Equipped with new knowledge of our genetic makeup, medical practitioners can aspire to meet challenges that recently appeared overwhelming. Even if doctors cannot yet treat the genetic disorders that affect some of the boys and girls of Children's Hospital—indeed, even though our abilities to treat *any* genetic disorders are still profoundly limited—genetic tests are rapidly becoming available, enabling pregnant women to discover whether the fetus they are carrying is afflicted with any of a growing number of severely incapacitating conditions. Enthusiastic champions of molecular genetics envisage a world in which fewer children lie inert among the stuffed animals, a world in which fewer parents observe, with unending anguish, the half-lives their children lead. Scientific knowledge will not eliminate the tragedies of Children's Convalescent Hospital, but by reducing their number, it can soften the edges of human suffering.

Others think differently. Critics of the rush to molecular medicine warn us not to think of genetic tests as part of a world in which doctors and parents work together to avert lives destined to be cramped and diminished. Instead, they fear that genetic testing will become another instrument for oppressing the many people who are already struggling with the inequalities of our affluent society. Turning away from the small numbers of tragic afflictions that would be prevented by prenatal testing for devastating genetic conditions, they invite us to consider the impact genetic information might have on the everyday lives of those whose lot is already difficult. They offer a different image.

It might be a small town dominated by large industrial buildings. Almost all those who live in the nondescript houses work at "the plant," earn "decent money," and lighten the monotony of their routine with daily banter. Often their jokes center on the risks of the job, risks considered unfit as topics of serious conversation. Their talk of "tickets to the graveyard" rings in loud contrast to the whispers about people who used to work there but who now lie surrounded by different kinds of machinery, in the county hospital. Sometimes mordant humor turns to anger, and militant representatives confront employers with demands for a cleaner, less hazardous workplace. The replies are grave, sympathetic, adamant: Everything that can be done is being done, to go any further would bankrupt the company, all the jobs would be lost, the management regrets . . .

One of the workers is a divorced mother who has grown up in this town, whose relatives live within walking distance. Thanks to their support, she can work full time, bringing home enough money to offer her children modest comforts—they have escaped the roughest, most run-down part of town, they are well fed, they have warm winter clothes. She can't afford luxuries, but she is proud of her independence, and although she knows the risks of working at the plant, she tries to forget the friend who died two summers ago. She reminds herself that she is young, that there will come a time when the children are grown and she will be able to find better work. She assures herself that cancer does not strike *that many* after all.

Her future, however, may be less bright. Biomedical research discloses that people bearing a certain *allele* (the form of a gene) are affected by the substances released in the plant and these people have a significantly higher chance of developing cancer: 15 percent instead of the 0.1 percent risk of those who do not carry the allele. Company executives have discussed the matter at length, commissioned detailed analyses of costs and benefits, and ultimately decided that it would be worthwhile to administer genetic tests to all the workers and release those who carry the unlucky gene. Among those who test positive is the single mother, and she receives notice that she will be fired. Requesting an interview with the management, she presents her case: No alternative work is available for her in this town; to move would destroy her system of support, the system that makes it possible for her to work full-time at all; she knows she is risking her health, but she must do so for the sake of her family. The management regrets . . . but the decision, she must understand, is in her own best interest.

How should we assess the likely consequences of the genetic revolution? Should we recall the children whose lives are violated by their terrible inheritance and focus on the good of preventing more lives like theirs? Or should we think of the families whose lives will be disrupted by the disclosure to employers or insurance companies of information about their unlucky genes? Both, surely, and more besides.

For there are many other prospects. Imagine the medicine of the twenty-first century as it might be. Advances in molecular genetics have disclosed the causes of numerous diseases. Clever molecular techniques make it possible to detect and eliminate many types of cancer cells. Hereditary predispositions to cardiovascular disease can be discovered very early in life, and frequently future problems can be avoided. Understanding of the molecular mechanisms of viral replication is so far advanced that entire classes of infectious agents—including HIV, "the AIDS virus"— can be subdued. Some of our descendants live longer, all enjoy healthier, more vigorous lives.

But are they happier? There is yet another image, the darker side of the optimism about medical progress. Perhaps in probing the mechanics of our lives, we will disclose differences that would better have been left hidden. Will the society of our descendants embody a new class system, one that distinguishes people on the basis of their genes, genes that are supposed, rightly or wrongly, to affect traits that are most prized? Will they try to plan future generations, designing combinations of people with combinations of genes so as to form a harmonious society? Will human life be reduced to a product, something whose quality is carefully monitored and controlled, made by licensed manufacturers under expert medical supervision? Dimly, fearfully, thoughtful people glimpse enormous successors to Baron Frankenstein's laboratory, twenty-first-century hospitals equipped with the "decanting rooms" of *Brave New World*.

Alternately inspiring and appalling, kaleidoscopic images of possible futures whirl by. We sense that the molecular revolution will make large differences—how large, we do not know—in the lives our children will lead, we sense that we have the power now to channel the impact the new biology will have on society, but the kaleidoscope shifts too quickly. We do not know how to stop it, how to bring these images into focus, how to decide which of them represents something for which we should genuinely hope or of which we have reason to be afraid.

My aim in the chapters that follow is to sharpen our pictures of possible futures. Which scenarios are based on fact, which on fantasy? What can be done to escape the harms many people rightly fear? How can we achieve the benefits for which many hope?

I shall begin, in chapters 2 and 3, with the most immediate practical implications of the new genetics. Within the next few years, our ability to test people and to determine whether they carry genes that place them at risk for various diseases or disabilities will be greatly multiplied. The power to diagnose will often coexist with scant improvements in treatment—indeed sometimes the test will simply reveal a future condition that medicine can do nothing to avert. Exactly what benefits is genetic testing likely to bring? What are the attendant difficulties? How do we maximize the former and minimize the latter?

Eventually, of course, we hope that increased understanding of the detailed processes that underlie diseases will translate into treatments, cures, and preventive measures. Inspired by the rapid pace of discoveries in human molecular genetics, it is easy to think that the golden age is just around the corner, that biomedical researchers have a systematic way to tackle diseases and disabilities that afflict millions. Chapters 4 and 5 will consider the ways in which molecular genetics is currently developing and the likely payoffs for the medicine of the next century. They attempt to deflate the most optimistic visions—there is no royal road to the universal health of future generations—while recognizing the power of our molecular methods to transform the treatment of some diseases in a hodgepodge of ways that are presently unpredictable.

We shall then consider the ways in which people in the future might be harmed by the collection of information about their hereditary material. How should we prevent new forms of discrimination from arising as those with unlucky genes find themselves barred from jobs or from obtaining health insurance (or other kinds of insurance)? Will those who have been victimized in the past discover that the new understanding of human genetics provides further ways of oppressing them? These questions, examined in chapter 6, are already forced on us: American families with histories of diseases such as cystic fibrosis have already discovered that they cannot secure insurance to cover treatments that would prolong their children's lives. As testing unveils the genetic material of many others, the questions will arise on a large scale.

Chapter 7 takes up issues that have acquired great prominence because of the potential role of DNA evidence in much-publicized trials. Only those resolutely determined not to watch television, listen to the radio, or to read newspapers and magazines could have avoided learning that there is a method of using DNA to convict the guilty, and that there are enormous controversies about whether the method has been properly applied. I shall try to explain just what is involved in these debates and offer some suggestions about how molecular genetics might both provide powerful forensic tools and also allow us to protect the rights of the innocent.

Beyond the immediate impact of the new molecular genetics is a set of broader questions that some will see as the most disturbing implications of the scientific advances. Our descendants will be able to discover all kinds of things about the alleles carried by fetuses. They will be able to use that information to choose, at least to some extent, the kinds of people who are likely to be brought into being. Are we in the process of beginning a profoundly immoral venture or of reverting to the evils of the eugenic past?

Once the question has been raised, we are plunged into many controversial issues. Can we separate the fight against disease from the attempt to shape society according to current prejudices? Should we use abortion as an instrument for promoting the health of our descendants? What will molecular genetics show us about the hereditary basis of human behavior? Will we discover that tendencies to traits considered socially undesirable—a propensity for violence, for example—are genetically determined, and so attempt to forestall the births of sociopaths? How will the knowledge we gain about the mechanisms of ourselves and our behavior be reconciled with our conception of what it is to be human? Will scientific self-understanding inevitably cheapen our lives? The second half of this book, chapters 8 through 14, is an attempt to explore and wrestle with these large questions.

The molecular revolution will change the ways in which people live, think, and act in profound ways: Human molecular genetics has implications that are immediate, practical, and concrete, and others that are subtle, abstract, and "philosophical." To come to terms with these implications it is important to acquire some understanding of the scientific basis of that revolution. In consequence, I have included in chapter 2 some details about the molecular discoveries of the past forty years and in chapter 4 an account of how molecular genetics is currently developing. Readers may prefer to treat chapter 2 as an introduction to some basic points about genetic testing and to a few prominent medical examples, skimming over the exposition of the science. Those who do so will have little difficulty in understanding some of the questions discussed later—they will have no trouble with the investigation of genetic discrimination (chapter 6) or with most of the topics of the second half of the book (the exception is the treatment of genetic determinism in chapter 11); I trust that they will also be able to appreciate the *main* points about genetic testing (chapter 3), about the likely future of molecular medicine (chapter 5), and about uses of DNA evidence in the courtroom (chapter 7).

Chapter 4, focused on the likely results of current investigations in molecular genetics, plays a special role in the argument of this book. At present, the United States government has committed $3 billion (spread over fifteen years) to the Human Genome Project. Enthusiasts have advertised this project by suggesting that sequencing the human *genome* (the totality of the genetic material found in a "typical" human cell) will lead fairly directly to great medical advances. A few naysayers have pointed out, correctly I believe, that there is no *direct* route from the biological information which is the official aim to the promised medical utopia. In chapter 4 I try to show why the current investigations are nonetheless valuable, why the real importance of the Project lies in apparently mundane investigations on organisms much humbler than ourselves, and why we can hope for an unpredictable potpourri of useful results. These conclusions form the basis for my assessment of the therapeutic promise of molecular medicine (chapter 5). Chapter 4 can be omitted by those for whom the details are not of central concern, but they will then have to take some parts of chapter 5 on trust.

Contemporary molecular genetics is a complicated but richly fascinating subject. I hope that many readers will find that my purely scientific discussions provide an entrée into one of the great achievements of human civilization, the fathoming of the intricate details of heredity. Yet this book is for those who are primarily concerned with the moral dilemmas and social problems that come in the wake of the great advances. If the capsule presentations of chapters 2 and 4 seem like whirlwind tours of seven capitals in five days, I can only apologize and urge them to visit the high spots of chapter 2 and to bypass chapter 4 entirely. Even if the nuances of some discussions are lost, the main themes should be clear enough.

As I hope to demonstrate in the chapters that follow, the topics surrounding human molecular genetics are diverse and intricate. Only an author of unconscionable arrogance could pretend to offer more than clarification of these complex issues and some tentative solutions to be taken as invitations for further serious reflection and debate. Yet one point should be firmly stated: Without reflection and exchange of ideas, we shall surely lurch into the future blindly; while forethought does not guarantee that we shall do better, it surely raises the chances that we shall avoid the deepest pits.

Human suffering takes many guises. Science is sometimes mercifully destructive, wiping away an entire species of familiar forms, sometimes cruelly creative, introducing new kinds. At the end of Friedrich Dürrenmatt's play *The Physicists*, a character who has pretended to be afflicted with the delusion that he is Einstein resumes his persona and remarks: "I love my fellow men, and I love my violin, and it was on my recommendation that they built the atomic bomb." Working under conditions of war, physicists in the early 1940s had little time for moral reflection. Today, contemplating molecular biology—a science that has at least comparable potential for impact on human lives—we have the luxury of time for thought. Mindful of the children of Children's, we can hope that not all their sufferings, and the grief of their families, need be repeated. Yet we should also remember the other images and beware lest, in our eagerness to stifle one source of anguish, we produce other terrible forms of human suffering.

2

Our Mortal Coils

The days preceding the baby's first checkup are full of tension. Of course, the new parents know they do not have to fear really grave news—they remember the terrible wait between the amniocentesis and the reassuring telephone call that explained that their child was free of major genetic disabilities. More fine-grained testing would await at birth. Just after delivery, the nurses took a sample of the baby's blood, and in accordance with the medical standards of 2020, the pediatrician will now give the parents the results of the analysis, the "genetic report card." The interview proves more frightening in prospect than in actuality. Although the long columns of statistics are initially baffling, the doctor points out that most of the risks are normal, or below normal, and that the only worrying figures are somewhat elevated probabilities of diabetes and hypertension. She recommends attention to diet from an early age, as well as regular checks of blood sugar levels. Like many others who have experienced the new medicine, the parents are grateful that they can take rational steps to promote their child's health.

We do not have to wait until 2020 for genetic tests. They are already with us, the first fruits of the new molecular biology. Their benefits are evident when we contemplate perfect cases. Molecular biological analysis might inform you that you bear an allele (a particular form of a gene) placing you at high risk for a painful incurable disease. The test you have taken is perfectly decisive: There is no mistaking the presence or absence of the allele, and if the allele is present, the probability of acquiring the disease is high; indeed, you are certain to be stricken unless you make some adjustments to your lifestyle. Luckily, you can make those adjustments, and although they would prove unpleasant and dangerous to people without the allele, they are essential for you. Prior to the existence of the genetic test, rational treatment was impossible: The modified lifestyle is harmful for members of the general population, and the unmodified lifestyle is lethal for bearers of the allele. Once the test became available, we could distinguish the carriers, and members of both groups were able to live happily ever after.

By 2020 tests for hundreds—if not thousands—of different genetic conditions will probably be available. Few of these will share all the features that make the perfect test so attractive. In many cases, perhaps even a majority, there will be a gap of uncertain duration between the development of a genetic test and the ability either to treat the corresponding disease or to reduce the risks of acquiring it. Intense research in molecular biology may yield quick results in some instances but reach a succession of dead ends in others. If a world of tests without therapies appears disconcerting, the best remedy may be to double our research efforts: Like Macbeth, we are already "in so far, that, should we wade no more / Returning were as tedious as go o'er."

Large-scale abnormalities in human genetic material—most famously, the presence of an extra copy of chromosome 21, which causes Down syndrome—have been detectable for decades. During the 1980s advances in molecular genetics led to a proliferation of genetic tests, as it became possible to discover if a member of a family with a history of hereditary disease carried an unlucky combination of alleles. The method was informative but not ideal: It required the cooperation of family members in supplying information and blood samples and allowed for both false positives and false negatives, although typically at low rates.

Recent breakthroughs in molecular genetics have changed the picture. More powerful techniques are rapidly becoming available, enabling doctors to discover the precise form of the genetic material that a person carries. In the next few years biomedical technology companies are likely to flood physicians' mailboxes with advertisements for genetic tests that can be given to anyone, whether or not the disease is known to run in the family, whether or not there are relatives willing to cooperate. Indeed, we shall be able to test at will—fetuses, couples who plan to marry, people who are sick, and those who show no symptoms.

Genetic testing may be valuable in extending diagnostic techniques, helping us to pin down the causes of ambiguous symptoms (as, for example, those for multiple sclerosis or elevated levels of cholesterol). But the tests that excite most attention are those that will allow physicians to predict some future state: the characteristics of the child who will develop from a fetus, or of the offspring of a couple who have not yet conceived, as well as the chances that a disease will strike someone later in life. So, with increasing frequency, doctors are taking blood (or cheek swabs) from patients who seem to be in excellent health, or from couples who are at risk of carrying recessive alleles for genetic diseases (such as Tay-Sachs disease among Jews of Ashkenazi descent, and thalassemia among Cypriots and other Mediterranean people). Similarly, prenatal testing is becoming ever more common. *Amniocentesis* (in which amniotic fluid containing fetal cells is removed, usually around the sixteenth to eighteenth week of gestation, sometimes earlier), *chorionic villi sampling* (a less familiar technique involving the removal of cells from the fetal part of the placenta, usually after the eighth or ninth week), and the extraction of cells from early embryos (in cases of *in vitro* fertilization) are all established medical techniques used to identify a growing number of genetic and chromosomal conditions.

• • •

To explore the potential benefits and disadvantages of genetic tests, those now available and those likely in the near future, it is important to step back and to bring into clearer focus the great achievements of late-twentieth-century molecular biology, the advances that have made possible the new cornucopia. Only in this way can we really understand how the tests work, what exactly they tell us, and what limitations they may have.

For nearly one hundred years biologists have known that the hereditary material—the genes—are segments of chromosomes, structures within the cell that are readily recognized under the microscope. The chromosomes of sexually reproducing organisms come in pairs, and when the organisms form their sex cells (or *gametes*)—sperm in males and eggs in females—each gamete receives one chromosome from each pair. Human beings have twenty-three pairs of chromosomes, so a human cell that is not a gamete (a so-called *somatic* cell) contains forty-six chromosomes, while gametes have only twenty-three each. When sperm and egg unite to form a fertilized egg (or *zygote*), the typical number of chromosomes, forty-six, is restored.

The process of transmitting chromosomes to gametes can sometimes go awry, producing a sperm or an egg with too many chromosomes or too few. When this happens, any fertilized egg that results will not have the typical number of chromosomes. Frequently, the effect will be lethal, and early miscarriages, while sad, sometimes should be viewed as nature's way of correcting some really bad mistakes, the gross errors of chromosomal assortment. However, having an extra copy of one of the chromosomes does not always terminate a pregnancy. Babies are sometimes born with extra copies of chromosomes 13, 18, and 21, as well as with unusual combinations of the sex chromosomes (the X and Y chromosomes; normal females are XX, normal males are XY). The consequent disruption of development is extremely severe when a third copy of chromosome 13 or of chromosome 18 is present. Mercifully, few of these infants survive for more than a year. Down syndrome, which results from an extra copy of chromosome 21 (sometimes an extra copy of only a part of chromosome 21), has a less profound impact.

Children with Down syndrome are always retarded to some degree, and many are quite severely retarded. Although some die young from malformations of the heart, most Down syndrome children acquire the rudiments of language, and most are able to walk. Because of efforts made in the past decades to nurture and challenge Down syndrome children, some have reached goals that used to seem quite unattainable, but it remains true that many children with the syndrome do not show marked improvements, even when reared in what we think of as the most supportive environments.

Besides the large-scale imperfections of inheritance reflected in the transmission of extra chromosomes, there are many minute accidents in distributing genetic material that can wreak equal or even greater devastation—or that can produce perfectly benign variations that help to account for the marvelous diversity of people. Genes occur at distinctive positions (called *loci*) along chromosomes. If we take any of your chromosome pairs, say your two copies of chromosome 19, then there will be a definite order of loci common to both chromosomes, and the forms of the genetic material you have at a particular locus will affect the trait (or complex of traits) associated with that locus. For example, molecular geneticists currently believe that they have identified a region of chromosome 19 that contains a number of olfactory loci. Consider any one of these loci. You have two chunks of genetic material occurring at this locus, one on each of your copies of chromosome 19. These are the two alleles you carry at that locus—the alleles at a locus are just the particular forms of genetic material that occur there. The precise character of these alleles will influence your sensitivity to particular kinds of smell.

"Influence" is a vague word, but it is exactly the right word in this context. How keen your nose is, whether you astound your friends with your ability to recognize the vintage (or even the vineyard) of the wine they are drinking, or whether you can run through urban pollution without a sniffle of discontent depends not on a single locus but on the combination of genes you have at many loci. Nor does it depend on the genes alone. Together the *genotype* (the combination of alleles you carry at all your loci), the internal composition of the zygote (the fertilized egg) from which you grew, and the environment in which you developed determine your manifest characteristics, your *phenotype*.

Strictly speaking, organisms do not inherit their observable traits. People do not pass on their wavy hair or fleshy noses. Genes are the links across the generations, and those genes may express themselves differently when they keep different genetic company or are housed in bodies that encounter novel environments. Aspiring gardeners everywhere should understand the point: No matter how much you invest in seeds or stock to secure high-quality genes, the vigor of your plants will be greatly affected by soil, sun, rain, and pests.

Nevertheless, in some cases, if a particular combination of alleles is transmitted, we know no way to reshape the environment to avoid a devastating effect. Apparently minor abnormalities in the form of the genetic material can doom a child of healthy parents to neural degeneration and early death.

Jews of Ashkenazi descent, as well as French Canadians, are at relatively high risk for Tay-Sachs disease. The course of this disease is invariable, although with extraordinary devotion parents can sometimes postpone the inevitable end for two or three years. At birth, Tay-Sachs babies seem quite normal, but during the first year their nervous systems degenerate, so that keeping them alive beyond the second year requires exquisite care in managing the most basic processes (such as feeding). Medical textbooks underscore the inevitability of the decline by dropping all standard qualifications: Under "Prognosis" we find the stark announcement "Lethal by 3 to 4 years."

Tay-Sachs disease is caused by the presence of two copies of an abnormal allele (a *mutant* allele, or *mutation*) at a locus on one of our chromosome pairs. Each parent carries one of the abnormal alleles, but each also has the normal allele at the locus on the other member of the chromosome pair, and that is enough for their normal development. Unluckily, in the formation of the gametes (the sperm and the egg) that fused in the zygote that became the child, both parents transmitted the abnormal allele, and the double dose is deadly. In the technical jargon of classical genetics, Tay-Sachs is caused by a *recessive* allele, an allele that hides its terrible effects when it occurs with the normal allele (in so-called carriers).

The typical differences between the mutant allele and the normal allele are tiny. Here there is none of the gross modification of hereditary transmission that occurs when an extra chromosome is transmitted (as with Down syndrome). How does a minute difference produce such enormous effects? To answer this question, we must venture further into the territory explored by late-twentieth-century biology, probing the structure of the genetic material.

In 1953 Britain celebrated the coronation of a new sovereign and the first ascent of Everest. Far less publicity attended a far more momentous event, the publication in the prestigious scientific journal *Nature* of a short article proposing "A Structure for Deoxyribose Nucleic Acid." Building on earlier but not widely appreciated work that had identified DNA as the genetic material, the two young authors, James Watson and Francis Crick, proposed the famous double-helix model of DNA, a model that would open the way to unanticipated biological and biomedical discoveries.

DNA molecules are composed of two backbones (made up of sugars and phosphates) that spiral around each other. Projecting inward from the backbones, like the rungs of a ladder, are the distinctive DNA subunits, four *bases* (or *nucleotides*): adenine (A), cytosine (C), guanine (G), and thymine (T) (see Figure 2.1). The two strands of the DNA molecule are complementary: If a particular base juts inward at one place on one backbone, then the appropriate counterpart projects toward it from the corresponding place on the other backbone;

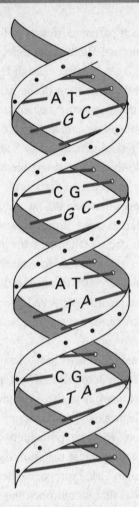

FIGURE 2.1. *The Structure of DNA*

DNA molecules consist of a pair of sugar-phosphate backbones wound around each other in the form of the familiar double helix. The bases, or nucleotides, protrude inward, like the rungs of a ladder. The order in which the bases occur on a single strand of DNA varies from DNA molecule to DNA molecule. However, there are strict rules of pairing. If cytosine (C) occurs on one strand then guanine (G) must appear opposite on the complementary strand, and vice versa. Similarly, adenine (A) and thymine (T) are always paired together.

adenine always pairs with thymine, cytosine always pairs with guanine. At one stroke, Watson and Crick had explained why, although the ratio of A to C (and G to T) varies from DNA molecule to DNA molecule, the ratios of A to T and C to G are always one to one.

Before the molecular structure of the genetic material had been fathomed, nobody had known whether uncovering the structure would illuminate gene functioning or would simply leave geneticists scratching their heads. The double helix was suggestive beyond all expectations. Even the conventions of sober scientific prose could not cover the excitement of future advances already glimpsed in the proposed structure. Watson and Crick concluded their article with one of the great laconic sentences in the history of science: "It has not escaped our notice that the specific pairing we have postulated immediately suggests a possible copying mechanism for the genetic material."

Getting copied is one of the principal tasks of the genetic material. Throughout our lives, the cells of our bodies (like those of many other multicellular organisms) are constantly dividing. The overwhelming majority are *mitotic* divisions, in which the descendant cells have the same number of chromosomes as the originals. (There are also, as described above, *meiotic* divisions, producing gametes, in which the chromosome number is halved.) Although mistakes in copying occasionally occur, the rule is for a single cell to produce two new cells, each with faithful copies of the DNA of the source. How does this work? Watson and Crick foresaw the elegant explanation. The two strands of the helix unwind, each serving as the template for the formation of a new strand that will be complementary to it. Because of the pairing of the bases—A with T, C with G—any double-stranded DNA molecule formed by attaching a new strand to one of the original single strands must be just like the two-stranded molecule with which the process began (see Figure 2.2).

Genes play important roles in two processes. They are transmitted from parents to offspring (a process that depends on their ability to be copied faithfully), and they help to cause the observable traits (the

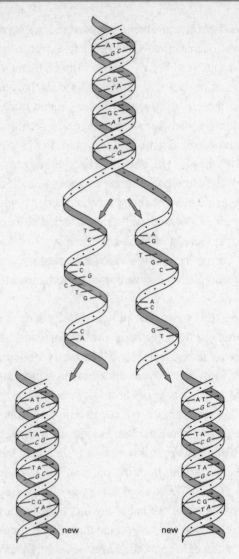

FIGURE 2.2. *DNA Replication*

When a cell divides to form two new cells, each of the descendant cells needs the full complement of genetic material. So cell division depends on DNA's ability to be copied faithfully. As double-stranded DNA is unwound, it is possible for new strands to be formed alongside the originals. Because of the fidelity of base pairing—A always going with T, C always with G—the double-stranded DNA that eventually results will resemble the original.

phenotype) of the organisms that bear them. After Watson and Crick, it was clear that, far from being boring, DNA molecules differed in ways that might easily affect the character of the phenotype. All contain the same four bases, but the order in which the bases occur along the strands is highly variable. How does the exact sequence of bases influence the characteristics of the organism? Within little more than a decade, molecular geneticists had discovered a major part of the answer by exposing the mechanisms of immediate gene action.

Thanks to some ingenious experiments on microorganisms (molds, bacteria, and viruses), it was already apparent by the mid-1940s that genes play a role in directing the formation of proteins inside cells. All the major processes that occur in plants and animals—breathing, digesting, or moving, for example—come about as the result of changes in the properties of numerous cells. The individual cells can do their special jobs because of the special proteins they contain. Like a complex enterprise that draws on the contributions of many factories, an organism copes with the world by coordinating the activities of its cells. The proteins are the workforce, and it is crucial to the organism's success that the right proteins are available. Genes can be thought of as the training staff who turn raw newcomers into workers with the special skills required to perform their assigned tasks.

Proteins are composed of amino acids linked together in a line and then folded up, often in extremely complicated ways. Many proteins serve as *enzymes* (molecules that catalyze chemical reactions inside a cell), and indeed, the connection between genes and proteins was established by showing that mutant organisms bearing abnormal alleles at a particular locus were unable to carry out vital reactions when placed in special environments. A major task for molecular geneticists in the 1950s and 1960s was to elucidate the connection between genes and proteins.

The story turned out to be more intricate than they had initially suspected. Genes play a role in forming proteins only when they are activated (when a special enzyme becomes attached to the DNA).

When a gene is *switched on* or *expressed*, it directs, through the activity of proteins already present in the cell, the formation of a complementary strand of *ribonucleic acid (RNA)*. RNA molecules are similar to DNA, except that they typically have one sugar-phosphate backbone instead of two, and that they contain uracil (U) in place of thymine (T). The newly formed RNA, *messenger RNA*, or *mRNA*, moves to a site in the cell at which proteins are formed. The process of protein synthesis depends crucially on proteins and other RNAs already present in the cell, but the sequence of amino acids in the new protein corresponds to the order of the bases in the mRNA. In other words, DNA is *transcribed* into mRNA, and the order of bases in the DNA determines the order of the bases in the mRNA; mRNA is *translated* into protein, and the order of bases in the mRNA determines the order of the amino acids in the protein. So in two steps, the sequence of the bases in DNA fixes the sequence of amino acids in the protein (see Figure 2.3).

Somehow the exact sequence of bases, the A's, C's, G's, and T's, in a gene affects the character of the protein transcribed and translated from it—but how? The molecular geneticists of the 1950s foresaw the existence of a "genetic code," a scheme according to which particular sequences of bases correspond to specific amino acids. In the 1960s they cracked the code. Bases are read in threes, each triplet (or *codon*) coding for a particular amino acid or signaling termination of the growing protein chain. Twenty amino acids are found in living organisms, and there are sixty-four distinct codons (because there are four possibilities—A,C,G,T—at each of three places). The genetic code is redundant (or *degenerate*): Six codons code for the amino acid leucine, four codons code for valine, and there are three *stop codons* (triplets that signal the end of transcription).

Cracking the genetic code solved many old problems. In the premolecular days, classical geneticists had puzzled over the nature of mutations. Presumably the organisms with abnormal characteristics, on which genetic analyses typically depended, had undergone some change in their genetic material that deprived them of the ability to

TRANSCRIPTION

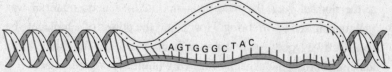

A G T G G G C T A C

Double-stranded DNA is
partially unwound

Newly formed RNA strand—
messenger RNA (mRNA)

U C A C C C G A U G
A G T G G G C T A C

DNA strand

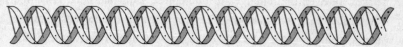

DNA is rewound

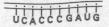

U C A C C C G A U G

messenger RNA is released

TRANSLATION

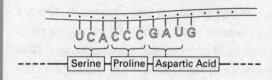

U C A C C C G A U G

Serine — Proline — Aspartic Acid

A chain of amino acids is
formed, making a protein.
The particular amino acids
that occur are determined
by the sequence of bases in
the mRNA, according to the
genetic code.

FIGURE 2.3. *Transcription and Translation*

When a gene is transcribed, an enzyme associates with the double-stranded
DNA that becomes partially unwound (*top*). A growing strand of RNA
forms, complementary to the exposed DNA strand. When transcription
ends at a triplet of bases that signals "stop" (a so-called stop codon), the
DNA is completely rewound and the messenger RNA (mRNA) is released.
The messenger RNA is translated in a complicated process, as specific
amino acids are lined up with triplets of RNA bases. The chain of amino
acids constitutes a protein.

function normally. Yet despite its failure to influence the phenotype in the normal way, the modified—*mutated*—genetic material was still transmitted to offspring. How could the power of a molecule be so selectively perturbed, stripping it of one of its functions while leaving another, the capacity for being faithfully copied, intact?

Molecular geneticists were able to answer this question by recognizing three major kinds of mutations. Some are small, involving a single change in the sequence of bases (or nucleotides) along a DNA molecule, replacing a C by a T, for example. When such *point mutations* occur, the effect on the protein chain will be, at most, to substitute one amino acid for another. (Recall that the genetic code is degenerate, so that changes in nucleotide sequence need not affect the sequence of amino acids.) Proteins often have hundreds of amino acids strung together, so that changing one may make little difference; the resultant protein may function in just the way the original did. Or the difference may be crucial: A point mutation is responsible for the disease sickle-cell anemia.

Potentially far more damaging are *deletions* (in which some bases are lost) and *insertions* (in which nucleotides are added), for these may affect the grouping of bases during transcription, leading to a protein with a completely different sequence of amino acids beyond the point at which the loss or gain occurs (see Figure 2.4). Often a deletion or insertion will cause a stop codon to appear prematurely, leading to a truncated protein. However, none of these changes, great or small, affects the capacity of the DNA for self-replication. The two strands can still unwind, and whatever the sequence of bases on the single strands, they will serve as templates for faithful replicas of themselves.

Small insertions and deletions can be devastating. Doctors have discovered that some infants born with severe respiratory disease, babies who can be kept alive for only a few months through the use of mechanical ventilation, suffer because their cells manufacture only a mutant form of a crucial protein ("Surfactant B"). The change is apparently tiny: A single occurrence of C is replaced with the se-

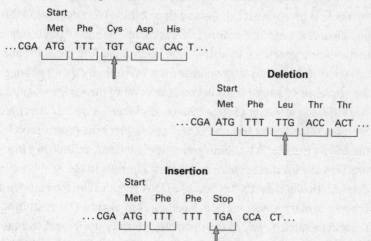

FIGURE 2.4. *Various Kinds of Mutation*

At the top of the diagram is a schematic picture of part of a gene, whose original unmodified sequence codes for a protein beginning with the amino acid methionine, two occurrences of phenylalanine, and so on. At the middle left, a single change—the replacement of a T by a G in the third codon—causes the substitution of one amino acid in the protein: Cysteine replaces the second phenylalanine, but otherwise the protein is unchanged. At the middle right, the deletion of one of the T's in the third codon produces a more dramatic shift: The third, fourth, and fifth amino acids in the protein are changed and, indeed, almost all the subsequent amino acids are likely to be different. A similarly dramatic alteration occurs at the bottom center, where the insertion of an extra T leads to the occurrence of a premature stop codon. In this case, instead of a protein with many amino acids, the result is a fragment with just three. Deletions and insertions, so-called frameshift mutations, are thus much more likely to make large differences to the structure and function of proteins than are single substitutions (or point mutations).

quence GAA. Inserting the two extra bases is enough to change the composition of the protein, to subvert the functioning of the lungs, and to make it impossible for the infants to breathe.

Similarly, one of the common mutations that leads to Tay-Sachs disease is produced by an insertion of four base pairs. Four base pairs hardly seem significant among the total DNA present within the nucleus of a human cell—some three billion base pairs—but this tiny change is crucial. Because DNA is "read" in units of three, when transcription generates an mRNA, the insertion of four bases (or one, or two, or five, or any other number not divisible by three) changes the sequence of amino acids beyond the point of the insertion, and in this case, the resultant protein cannot discharge its normal function. So long as the bearer has one gene that can yield the proper protein, there is no trouble. When both genes have mutated, cells do not manufacture any normal protein, and that lack results in the baby's early deterioration and death. Yet because DNA is copied by the pairing of complementary bases, carriers have an even chance of transmitting the mutant allele to any of their gametes. Besides the large disruption that four extra base pairs can bring, the mindless fidelity of DNA copying is part of the tragedy of Tay-Sachs.

How can doctors discover if someone carries a potentially dangerous allele, perhaps one that might be transmitted to a child or one that might presage elevated risks of disease for its bearer? Sometimes they begin with insight into some of the details about how the disease is caused. For example, they know that when a particular protein is missing or defective, there are serious physiological effects, and they know how to probe the body to discover whether normal or abnormal forms of the crucial protein are being produced. In these instances, they can test in advance of knowing the location and structure of the underlying gene(s): Tests for Tay-Sachs became available as soon as biomedical researchers were able reliably to pick out the mutant protein. Frequently, however, that strategy fails because of deep ignorance about how the disease works. Genetic analysis reveals that mutant alleles are being transmitted and that a particular combination of them produces debilitating disease, but what exactly goes wrong, what differences occur in the cellular chemistry and physiology of those who suffer, is totally unclear. Indeed, as the example of Surfactant B reveals, we may sometimes discover the physiological changes involved in the disease by exposing the structure of mutant alleles. Instead of knowing the protein first and then finding the gene, we discover the structures of normal and defective proteins by first locating a gene and determining the structures of its alleles. Massive ignorance is an omnipresent part of the medical condition. The great triumph of human molecular genetics in the last decade has been to show how this ignorance can be alleviated.

Until recently many major genetic disorders, including Huntington's disease and cystic fibrosis, resisted the kind of biochemical analysis required to develop a test. Huntington's disease, recognized since the nineteenth century, is a late-onset disorder that strikes both men and women, usually between the ages of thirty and fifty, and brings years of ravaging neural deterioration, physically (and surely mentally) painful to the sufferer and emotionally agonizing to those who watch the living death of a person they once knew and loved. Unlike Tay-Sachs, the Huntington's allele is dominant—a single dose is deadly. Any child of a parent who becomes afflicted with Huntington's has an even chance of having received the allele, and if the allele has indeed been transmitted, then the child too will develop the disease. Of course, because the disease strikes late, people may conceive a child without knowing at the time of conception whether they themselves will develop it. (They may even be ignorant of the risk if, for example, their own parent's condition was misdiagnosed or if that parent died by accident before the time of onset of the disease.) For those of us over forty, the poignancy of Huntington's was brought home by the portrayal of Woody Guthrie in the film *Alice's Restaurant*; more recently, the dedication and energy that Nancy Wexler (whose mother died of the disease) has brought to the search for the molecular causes has made the plight of those who suffer from Huntington's and the anguish of those who know they might come to suffer from it even more vivid.

Until the 1980s there was no method for knowing in advance which of those at risk for Huntington's would indeed contract it; no means, for example, of discovering the abnormal protein that bearers of the mutant allele produce. (Nobody knew anything about that protein or what its damaging effect might be.) Similar darkness surrounded cystic fibrosis, a condition that afflicts about one person in 2,500 (among people of Northern European origin) and occurs at much lower frequency in African Americans and in people of Asian descent. About a decade ago, patients with CF typically died in childhood or early adolescence. Some children develop complications of the pancreas and the intestine, but the most common effect of the disease is on the respiratory system; 90 percent of CF patients die from respiratory failure. Airways in the lungs become clogged with thick mucus, providing breeding grounds for infectious agents. The genetic entanglement was clear: CF results when two copies of a recessive allele are present. Yet nobody knew what the mutant alleles do, how they exert their damaging effects, or why they fail to provide something that healthy bodies need. Indeed, CF often proved difficult to diagnose and, in retrospect, we can see this as contributing to the contraction of lifespan.

Yesterday's ignorance has been dissipated. Within the past few years, the genes for Huntington's disease and cystic fibrosis have been mapped, cloned, and sequenced. "Mapping," "cloning," "sequencing"—these are the terms of the new trade in molecular genetics, and that trade has given us the power to test on a grand scale, to discover the structures of the normal and mutant alleles of a gene without knowing in advance anything at all about what those alleles do, what kinds of proteins they encode. How have such extraordinary advances in knowledge been possible? To understand the success and the promise, the scope and the limits of biomedical testing, we must acquaint ourselves with the arcane techniques of the molecular virtuosi.

• • •

From everything I have said so far, you might easily construct a picture of your genome, the totality of your genetic material. That picture would show twenty-three pairs of chromosomes (twenty-three long DNA molecules, collectively containing three billion base pairs' worth of DNA) tightly packed with genes, with each gene coding for a distinctive protein. An efficient Creator might have designed organisms with chromosomes chockablock with functional DNA, each gene neatly abutting its neighbor genes. However, living things are products of the inefficient, opportunistic processes of evolution, and in complex multicellular organisms (such as ourselves) the genes are scattered sparsely in the totality of the DNA (see Figure 2.5a). Between the genes lie vast intervening stretches containing sequences of A's, C's, G's, and T's that do not code for any protein.

Some parts of this intergenic DNA play a role in regulating the expression of genes. In multicellular organisms (ranging from plants, worms, and flies to people) the genetic material is virtually the same in most somatic cells (recall that these cells, unlike the gametes, have the full complement of chromosomes, and changes in the DNA occur only occasionally through copying errors). Muscle cells, neurons, skin cells, hair cells, liver cells, with all their distinctive properties, contain the same genes. If all the genes are the same, how do the cells come to be so different?

Since the 1960s molecular geneticists have known the general shape of the answer. Genes are transcribed at different rates in different types of cells. The characteristics of a cell, and ultimately the phenotype of the organism, depend on the selective switching on and switching off of genes. Pieces of DNA sequence outside (and sometimes within) a gene affect the conditions under which that gene is transcribed. If an unlucky mutation occurs in a regulatory sequence, the gene responsible for producing a necessary protein may never be transcribed, and the effect may be as severe as if a change had occurred within the gene itself.

Regulatory regions are often sites adjacent to a gene to which

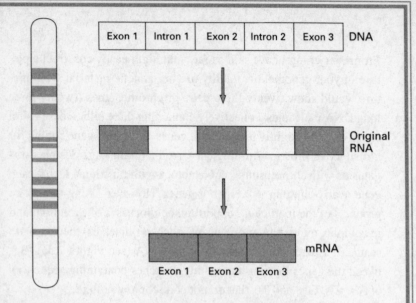

Figure 2.5a: Schematic illustration of the distribution of genes on a chromosome. The shaded areas correspond to genes; the unshaded regions

Figure 2.5b: Schematic illustration of a gene in organisms with a nucleus (eukaryotes—a wide range of organisms that includes yeast, worms, insects, plants, fish, birds, and mammals).

F I G U R E S 2.5a and 2.5b. *Schematic Structure of a Chromosome and of a Eukaryotic Gene*

At the left (a) is a schematic picture of a chromosome on which a few genes are scattered with large stretches of noncoding DNA between them. Typical human chromosomes have many more genes, but show something like this proportion of genes to intergenic DNA.

At the right (b) is a schematic illustration of a eukaryotic gene (*eukaryotes* are organisms whose chromosomes are found within a cell nucleus), showing how the segments of RNA transcribed from the introns are excised before the final messenger RNA is formed. In actual cases there may be many more exons and introns, and the relative sizes of exons and introns may vary considerably.

molecules in the cell can attach, making the DNA accessible or in-accessible to the enzymes that start transcription (thus beginning the first step in the building of a protein). Although we do not yet know how much of the intergenic DNA serves a regulatory function, there seem to be vast stretches of "junk" DNA that appear to have no func-tion. Parts of these DNA wastelands resemble modified copies of genes. Others iterate the same sequences mindlessly for thousands of bases: Sometimes the repeats are built on a short unit (. . . CA-CACACA . . .); sometimes on longer periods. It is as if our total DNA were a peculiar book in which an author addicted both to revi-sions and to doodling has bound all the early drafts, notes, and jot-tings along with the final version.

Nor are the passages that make coherent sense printed continu-ously. In organisms whose cells contain a nucleus (organisms rang-ing from yeasts to mammals), the original RNA transcribed from the chromosomal DNA is modified, and by the time the mRNA leaves the nucleus, particular pieces of sequence have been excised. Only selected parts of the continuous stretch of DNA that gave rise to the first RNA transcript are encoded in the sequence of amino acids con-stituting the final protein. These parts are the *exons*. The *intervening sequences*, or *introns*, correspond to those parts of the original RNA that are removed in making the mRNA, and which thus play no role in translation (see Figure 2.5b). Are these intervening sequences simply more junk? Not always. Even though those sequences are not represented in the final protein, molecular geneticists have discov-ered that if the introns are absent or are modified, the gene may fail to function normally.

Today human molecular geneticists have set for themselves the ambitious task of bringing order to the apparently ramshackle mess that is the totality of the genetic material in members of our species—they have already begun to map the human genome. Cer-tain kinds of changes in our genetic material cause us medical prob-lems, some of them extremely severe, and the aim is to locate the particular stretches of DNA that produce specific troubles. At first sight, the project looks quite hopeless: Genes are scattered at low density among three billion base pairs, they are divided by long stretches of molecular nonsense, and they contain within themselves segments sufficiently dispensable to be excised in the process of protein formation. Plainly virtuoso mapping techniques are needed.

To discover the locations of particular genes, researchers need reliable ways of picking out DNA segments and of making multiple copies of those segments. The first important step is to cut, paste, and copy pieces of DNA. As is so often the case, the trick is to turn natural processes to new ends, and in this instance, geneticists learned how to direct a family of processes that occur naturally in bacteria. Bacterial cells contain many molecules, *restriction enzymes*, which break down pieces of foreign DNA or RNA. Each restriction enzyme recognizes a specific sequence of nucleotides and cuts DNA at a particular site within (or near) its target sequence (see Figure 2.6). If a collection of restriction enzymes is let loose on some DNA, the enzymes will seek out their favorite targets, chopping up the DNA into a large number of fragments that end with the sequences at which the enzymes like to cut. Provided the fragments are the right size, they can be combined with the DNA of other organisms, using enzymes that function to glue pieces of DNA together. Borrowing the right molecules from the right bacteria, molecular geneticists have been able to play "cut and paste," bringing together new combinations of DNA, splicing genes they want to investigate into bacteria that multiply rapidly, and thereby generate many copies of the desired genes.

Moving pieces of DNA around would hardly be useful unless there were ways of knowing which cells or which containers hold the bits that are of interest. Although the DNA molecules that scientists manipulate are large—for molecules—they cannot be followed by zooming in with the light microscope. A host of ingenious techniques is needed to prevent them from going incognito. Most of these techniques depend on the rules of base pairing (A with T, C with G) that Watson and Crick discovered. DNA can easily be

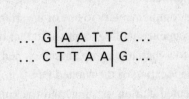

FIGURE 2.6. *Restriction Enzymes Cut DNA*

The diagram focuses on a small stretch in a much longer piece of double-stranded DNA. The small segment contains a recognition site for the enzyme EcoRI, which cuts the DNA as shown, thus producing two fragments with protruding ends.

induced to unwind, forming single strands. If geneticists now introduce a "probe," a piece of single-stranded DNA from the source that concerns them, neatly tagged with some radioactive atoms, then the probe will be complementary to one of the single strands they are out to locate (being complementary to one of the strands from the source means that its sequence will be identical to that of the other). It will bind to strands to which it is complementary, and the radioactive tag will disclose the location of the wanted DNA on X-ray film.

The sophisticated cutting, pasting, and tracking techniques of recombinant DNA have already proved their worth in medicine and agriculture. For well over half a century diabetics have been kept alive through daily injections of insulin extracted from the pancreases of cows and pigs. For some patients, human insulin proves more beneficial, and since the advent of molecular cloning, it has been possible to make human insulin on a grand scale. If an appropriate piece of human DNA is inserted into the useful bacterium *E. coli* and is accompanied by a regulatory sequence that switches on the gene coding for normal human insulin, then bacterial cells are transformed into insulin-producing factories. Subsequent ventures have yielded hormones (such as human growth hormone, which has been employed to treat some forms of dwarfism), vaccines (such as the vaccine for hepatitis), and blood-clotting factors (bringing relief to hemophiliacs). As techniques for cloning larger segments of DNA improve, medical researchers look forward to an enormous new arsenal of effective, safe drugs. In similar fashion, by inserting "foreign" DNA into the genetic material of plants, agricultural geneticists have produced strains resistant to disease or environmental stresses. Tomatoes bearing a gene that delays rotting enable harvesters to pick their crop later, and the resulting riper fruits have already titillated the tastebuds of Californian consumers. Visionaries anticipate a future in which fruits, vegetables, and grains are engineered to retard spoilage, so that the major obstacle to feeding the world's hungry, the problem of distribution, can be efficiently overcome.

• • •

Basic research is seldom accurately evaluated by those who crave instant gratification. Impressive as their recent successes may be, many molecular geneticists believe the true importance of recombinant DNA research lies in the *theoretical* advances it will make possible. By breaking our DNA into small pieces, each of which is copied many times, the stage is set for investigations of the *interesting* pieces. Genes are scattered on chromosomes, and chromosomes are essentially very long DNA molecules. Researchers hope to identify the parts of the DNA—the genes—that are responsible for the formation of proteins, especially in those cases where mutations cause medical troubles.

To realize that hope, molecular geneticists need maps. The enterprise of genetic mapping goes back to the dawn of classical genetics just before World War I, and to the brilliant insight of a second-year Columbia University undergraduate, Arthur Sturtevant, who recognized how biologists could take advantage of the quirks of the process in which gametes are formed to determine the order of genes along a chromosome.

Just prior to the division that yields sperm or ova (gametes), the chromosomes line up in pairs, and one member from each pair will be transmitted to each normal gamete. However, while the chromosomes are aligned, they can wrap around one another and exchange segments in a process called *recombination*. Alleles on a chromosome at loci that are very close together are likely to remain linked, whereas alleles that are further apart may be separated (see Figure 2.7). By looking at very large numbers of organisms and observing the frequencies at which the phenotypic traits associated with particular alleles are separated and are newly conjoined, it is possible to order the loci—the rarer a recombination, the more closely linked are the pertinent loci. Working with fruit flies, which breed rapidly and can be mated ("crossed") in enormous numbers, Sturtevant and his teacher, Thomas Hunt Morgan, were able to make the first genetic map, showing how some loci were associated with particular chromosomes and how they were distributed along those chromosomes.

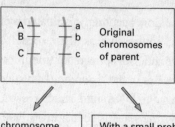

A ——
B ——
C ——

a ——
b ——
c ——

Original
chromosomes
of parent

Most offspring get a chromosome
with one of the original combinations
of the three alleles,
either   or

A ——
B ——
C ——

a ——
b ——
c ——

With a small probability there can be
a break between the **A** locus and the
C locus.

A —— a
c —— C

A —— a
c —— C

In these cases the offspring will
receive one of the recombinant
chromosomes
either   or

A ——
c ——

a ——
C ——

Because some but not all of
the organisms that receive the
new combinations **Ac** or **aC**
have the combinations **Ab**, **aB**,
organisms with the latter
combinations are now rarer
than those with the former. So
when we discover that the **Ab**
and **aB** recombinants are rarer
than the **Ac**, **aC** recombinants,
we can conclude that the **B**
locus is between the **A** locus
and the **C** locus.

In a fraction of these cases the break
will have occurred between the **A**
locus and the **B** locus, so that the
offspring will receive one of the
recombinant chromosomes
either   or

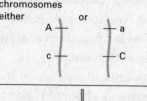

A ——
b ——
c ——

a ——
B ——
C ——

FIGURE 2.7. *Mapping Loci*

This is a schematic depiction of the technique for mapping loci pioneered
by Sturtevant and his colleagues. The trick is to look at many offspring
from parents who have different alleles at several loci and to use the frequencies with which various kinds of recombination occur to determine
the relative positions of loci.

Human beings cannot be bred quickly in large numbers, should not be bombarded with X rays to induce mutations, and should not be "crossed" to satisfy the experimentalist's curiosity. Without the standard tools of the geneticist, our knowledge of the organization of genes in our own species was haphazard until the early 1980s, when a group of scientists saw how to extend Sturtevant's insight. There is no need to manipulate people if we can make use of natural experiments, the everyday redistribution of variations in genetic material that goes on all the time as we reproduce. Tracking linkage and separation of significant genes is extremely hard, since our knowledge of the genetic basis of human characteristics that we can recognize is so limited: It is just our ignorance in this area that we are out to amend. However, if we focus directly on human DNA, detecting variations in the sequences of bases in a particular region, we can examine the ways in which those variations are transmitted, studying which ones go together most of the time, and using these kinds of differences just as Sturtevant and Morgan used the observable, heritable characteristics of flies.

How then is it possible to "focus directly" on differences in human DNA? Once again, those helpful bacterial molecules that chop up DNA come to the rescue. Suppose we take your two copies of chromosome 17 and my two copies and we break them up into small pieces by using some restriction enzymes. We can discover which pieces of your DNA correspond to which fragments from mine. Assume, then, that we know that the pieces labeled "A" and "B" come from a particular region on my two copies of chromosome 17, and that the pieces labeled "C" and "D" come from the same region on your two copies. If we now treat all these fragments with a new enzyme, one that we did not originally use in chopping up the chromosomes, it may have a target within a fragment or it may not. More interestingly, it may recognize a sequence within B, C, and D, but not within A: Perhaps one of my chromosomes has undergone a mutation so that the sequence at which the enzyme likes to cut has been altered. The result will be two fragments for B, C, and D, and only one fragment for A. Provided we can detect when we have produced two fragments rather than one, it will be possible to discern genetic differences between us.

These differences are examples of *restriction fragment length polymorphisms* (*RFLPs*) (see Figure 2.8). Roughly, RFLPs occur at points in the DNA where letting loose a well-chosen restriction enzyme enables geneticists to detect differences in the genetic material of different people (or, indeed, in the genetic material of organisms belonging to the same species). More exactly, a RFLP is a locus at which cutting DNA from different people (or organisms of the same species) will yield fragments that sometimes differ in length. RFLPs can be treated like alleles that undergo the simplest type of genetic transmission. Thousands have already been found in human beings, and by studying their occurrence in human genealogies, scientists can localize them to chromosomes and determine their relative positions. They serve as landmarks within the human genome, and once we have enough of them, we can use them to situate the genes. As we shall see shortly, this is not simply a matter of arranging genes that are already known, but, more significantly, of discovering new ones.

RFLPs can be studied only because there are ways of identifying the relative sizes of DNA molecules. The fundamental technique, *gel electrophoresis*, consists of allowing DNA molecules to migrate through an agarose gel matrix under the influence of an electric field. Because DNA molecules carry a negative electric charge, they move toward the positive pole. Smaller molecules move faster than large ones and thus go further in the same time. After allowing enough time for the molecules to separate, the current can be switched off, and with appropriate treatments, the molecular positions can be made to show up as bands on an X-ray film. So researchers can detect that one of the original sources of DNA, set into one column (or lane) in the gel, has produced a single large fragment, while the others have produced two short fragments that have migrated further.

They can also make more subtle comparisons. Even if my DNA does not differ from yours in the loss of a *restriction site* (one of

(a)

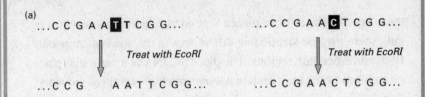

(b)

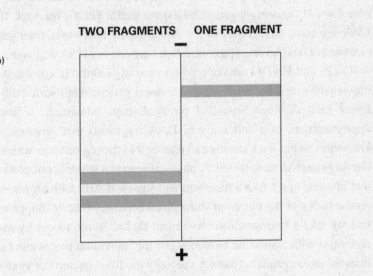

FIGURE 2.8. *RFLPs*

(a) Suppose that the two pieces of DNA shown have been obtained from different people. When these pieces are treated separately with the enzyme EcoRI, the one on the left will break into two fragments because it contains an occurrence of the sequence at which EcoRI cuts; the single base change in the DNA on the right has altered the sequence so that the site for cutting with EcoRI is no longer present. Thus the piece of DNA on the right will produce two fragments, the one on the right only one.

(b) When the fragments are placed in separate lanes of a gel, the differences will be revealed (provided that the fragments have been "labeled" by attaching a radioactive molecule or a molecule that fluoresces in a distinctive way). The two short fragments *(left-lane)* move farther down the gel in the time that the one long fragment *(right-lane)* takes to go a much shorter distance. So we can "see" that the two people differ in this region of their DNA, and even though the sequence doesn't contribute to their ordinary, observable properties, we can identify different "alleles" at a RFLP locus.

those special pieces of sequence that attract restriction enzymes to cut), there may be identifiable differences in the sizes of fragments from corresponding regions. The piece of DNA at a particular place on a chromosome may contain a long stretch of repetitive sequence. People often differ in the number of the repeats, and so-called *variable number tandem repeats* (*VNTRs*) are useful genetic markers. If you have sixty-nine copies of the repeat to my thirty-two, then gel electrophoresis can reveal the variation between my DNA and yours.

RFLPs and VNTRs are not only enormously helpful in constructing genetic maps, but they can also connect genetic maps with a different kind of map. So-called physical maps consist, to a first approximation, of a collection of DNA fragments that geneticists know how to put back together to reassemble the region from which the fragments were obtained. A physical map is a whole lot of physical objects, much like a disassembled jigsaw puzzle with numbers on the backs of the pieces to show just where they belong. Suppose that we take a human chromosome and shatter it into pieces by attacking it with restriction enzymes. All the individual pieces can be inserted into organisms (usually specially modified bacteria or yeast) that will generate plenty of copies, so we obtain a collection of clones, stored in a freezer in the laboratory basement. Each of the fragments can be labeled with genetic markers, RFLPs or VNTRs, whose locations on a genetic map are known. The ideal would be to put them all back in the right order, assigning them numbers that indicate just where they belong on the chromosome. If we had a rich enough genetic map connected to a complete physical map, we could pull out from our store a clone containing any genetic material of interest.

Presently, we are on the way to achieving that ideal, and for some regions of some human chromosomes, quite detailed genetic and physical maps have been constructed. The new maps make possible a powerful approach to human genetics. Starting from a pattern of hereditary transmission, researchers look for markers associated with the inherited characteristic. They discover, for example, that in a family with a history of a certain disease, a particular RFLP allele regularly accompanies the presence of the disease. When the DNA is chopped up and the appropriately treated fragments are allowed to migrate through a gel, the pattern of bands for those who suffer from the disease is different from the pattern for those who do not. With hard work—and luck—it is possible to find markers, RFLPs or VNTRs, that flank the locus implicated in the disease. Knowing that the gene they are hunting lies between two markers, geneticists can then use their physical map to pick out a small number of clones for intensive study. Using a variety of sophisticated techniques, they can "explore" the DNA in these clones, seeking out stretches that are likely candidates for genes, and, with luck and patience, they can eventually discover the locus at which mutations produce the disease. Markers make genes known by the company they keep.

This strategy, *positional cloning*, is responsible for the breakneck pace of recent discoveries of genes implicated in various diseases: Huntington's disease, cystic fibrosis, as well as some forms of breast cancer and colon cancer are prominent examples of success stories. In all these cases, researchers began in darkness. They had evidence that the diseases were inherited, but they had little or no idea about what the genes involved do or about the molecular processes that go awry. Good maps have enabled them to journey to understanding. Associating the disease with markers, they can discover just where the locus is. Once they know where it is, with luck they can find the structures both of the unproblematic alleles and of those mutations that produce diseases. At that point, they have the power to test, to analyze any of our DNA and to discover its structure, to determine whether a mutation implicated in disease is present.

In fact, even without knowledge of exactly where the crucial locus is and of the structure of "good" and "bad" alleles at that locus, it was already possible during the 1980s to test people whose family histories revealed them to be at risk for such diseases as Huntington's. Once you know that a particular genetic marker regularly accompanies the disease within a particular family, you can test family members who have not yet reached the age at which the disease typically strikes by looking for the pertinent marker. The marker is a good guide to the presence of the allele implicated in the disease. Good, but not perfect. Situated on the chromosome at some distance from the locus—perhaps one or two million base pairs away—the marker may become separated from its companion allele in the process of genetic recombination. So the RFLP tests of the 1980s, helpful though they were, were not conclusive; both false positives and false negatives could occur.

Looking for genetic markers can be helpful before the structure of normal and mutant alleles is known, but the tools of molecular genetics allow us to do better. Since the early 1970s biologists have learned to "read" the sequences of stretches of DNA. Like the reading performances of small children, the first efforts were slow and halting, restricted to small pieces of molecular text; today's sequencing is far more fluent and wide-ranging, although as we shall see later, it may still not be up to the task of decoding the three billion base pairs of the human genome. For purposes of genetic testing, sequencing is already fluent enough. Genes are typically measured in thousands of base pairs, and although the project of finding a locus may be long and arduous, generating the sequences of the "good" and "bad" alleles there is usually routine.

There are two main methods which can provide the sequence of A's, C's, G's, and T's that make up a chosen piece of DNA. Both rely on gel electrophoresis to separate DNA fragments that differ in length by only a single base. The more popular of these is the so-called *chain termination method*. Here the trick is to persuade a single strand of DNA, the strand whose sequence is sought, to build complementary strands of different lengths.

Imagine that you place single strands of DNA, all obtained from the same clone and therefore sharing the same sequence, in a solution containing the molecules that would normally, in a living cell, cause complementary strands to form. After a short while, the solution will be full of double-stranded DNA: The usual reactions will have occurred, and the original single strands will be paired with complementary strands that extend their full length. However, if some of the bases (the A's, C's, G's, and T's) are suitably modified (if they are so-called *dideoxy bases*), the complementary strands will vary in length. Dideoxy bases only cooperate partially in the formation of DNA strands: They can be taken up into a growing DNA chain, but they will not let the chain grow beyond them. So if you began by loading the solution with some normal bases and some modified bases (and if you have cleverly chosen the ratio of normal to modified bases), you will have a mixture of strands of varying lengths, including strands that are one base shorter than the original, two bases shorter than the original, and so forth.

Now imagine that you run the experiment four times. The first time you run it, all the C's, G's, and T's in the solution are normal, and some of the A's are modified. The modified A's have been radioactively tagged. When you separate all the resultant DNA into single strands, and run it through a gel, the fragments will be spread out. After the spread has been exposed to X-ray film, you will see the positions of a number of strands of various lengths, and each of these strands must terminate in a modified A. (Only the modified A bases have been radioactively tagged, so only the strands that terminate in these bases show up.) Suppose that your gel also contains three other lanes and that you run through each of these the DNA obtained from experiments in which the only modified bases are C's, G's, and T's, respectively, with the modified bases radioactively tagged in each case. Then the picture you obtain will reveal the shortest fragments at the bottom and the longest ones at the top. If the very shortest fragment is in the lane corresponding to the experiment in which the G bases were modified, then you know that the sequence (of the strand complementary to the original DNA) starts with G; if the next shortest is in the lane with the modified A bases, you know that the second base is A; proceeding from the bottom up, you can read the sequence of the complementary strand and, using the base pairing rule (A with T, C with G), you can derive the sequence of the DNA you originally obtained from the clone (see Figures 2.9a and 2.9b).

These days, people are not needed to do the reading. Thanks to advances in the 1980s, bases can now be tagged with fluorescent dyes (with different colors for each base), and the fragments can be separated in a single lane of a gel, revealing a distinctive array of colored bands that can be read by a machine. One lane now allows for sequencing a DNA segment of five hundred bases. Where, in the early 1970s, the pioneers of sequencing technology labored to produce a hundred or so reliable bases of sequence over a period of months, many laboratories can now generate a few thousand bases in a day.

Genetic maps give diagrammatic representations of the order of genes and genetic markers; physical maps tell us just how labeled DNA fragments fit together; the sequence, sometimes conceived as "the ultimate map," discloses the fine structure of a region of DNA. Deploying these tools in combination, molecular geneticists have transformed human medical genetics, making it possible to go from the bare recognition of hereditary transmission of disease to the isolation of alleles that cause disruptions of normal functioning. When some cases of a disease are produced by mutations at a single locus, developing a genetic test is a matter mostly of hard work and time. The genetic report card of 2020 could easily become reality, the natural extension of the research that is currently being done.

Tests are already available for a number of genetic diseases. Huntington's disease, we now know, results from an abnormally long trinucleotide repeat (CAG) in a dominant gene near the tip of chromosome 4. Unaffected people have between 11 and 34 repeats, those with Huntington's have more than 42, sometimes as many as 100. Although the age of onset is variable, it seems to be associated with the length of the repeat; the longer the repeat, the earlier the disease strikes. Hence, by analyzing the DNA of a person from a Huntington's family (or, indeed, anyone's DNA), it is possible to discover how many copies of the repeat are present, and thus whether the disease will develop. (The length of the repeat also allows a conjecture about the age of onset.)

Like Huntington's Disease, myotonic dystrophy is caused by an unusually long trinucleotide repeat within a dominant gene. When a person carries the long repeat, there is always some degree of progressive muscle weakness: Some children die at birth, others have little muscle tone from infancy on (and also suffer mental retardation), yet others develop almost normally to adulthood (and despite some degree of muscle degeneration, have children of their own). Here, too, it seems that the severity of the disease and its time of onset are associated with the length of the trinucleotide repeat.

The most common cause of mental retardation, Fragile X syndrome, has also been traced to an abnormal trinucleotide repeat. A gene on the X chromosome is interrupted by a number of copies of the sequence CGG. Normal members of the population have between 6 and 54 copies. Those affected with the syndrome typically have more than 300 copies, and may have as many as 1,500. Those with an intermediate number of copies are said to have the premutation: They do not usually display any symptoms, but there is a significant chance that they will transmit to their children an allele with an even larger number of repeats, thus causing the children to have the full syndrome. Women with very long repeats (more than 300 copies) are sometimes mentally handicapped (in about 20 to 30 percent of cases there is moderate retardation), and in the population so far tested, men are always mentally handicapped to some degree. Although IQ scores among Fragile X boys and men range from borderline retarded to severely retarded, almost all male patients suffer from behavioral problems. They frequently engage in repetitive speech, bite their hands, and show some of the emotional responses associated with autistic children (most notably a lack of appropriate reactions to the emotions expressed by others). Contemporary medical practice can do little to alleviate these problems, and many Frag-

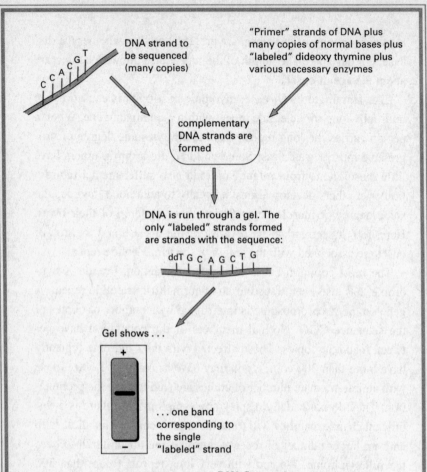

FIGURE 2.9. *DNA Sequencing*

(a) At the top of the diagram, many copies of the DNA strand to be sequenced are placed in a mixture containing normal bases, "labeled" dideoxy bases, and a number of other necessary molecules. Complementary strands of DNA are formed. Most of these are full of normal bases and run the whole length of the original strand of DNA. But some incorporate one of the dideoxy bases, so that the growing chain is terminated at that point. In the experiment shown, the chain is built from right to left, and there is only one occurrence of the complement of the dideoxy base (only one occurrence of A in the original strand, so only one place for a T to go in the complementary strand). So just one "labeled" strand will be formed. When the DNA is placed in a gel and the "label" is identified (by exposing the gel to X-ray film if the "label" is radioactive), there will be a single band.

Repeating the same procedure with different "labeled" dideoxy bases.

C C A C G T

| normal bases, etc., plus "labeled" dideoxy thymine | normal bases, etc., plus "labeled" dideoxy adenine | normal bases, etc., plus "labeled" dideoxy guanine | normal bases, etc., plus "labeled" dideoxy cytosine |

| ddT | ddA | ddG | ddC |

longest strand G
• G
• T
• G
• C
shortest strand A

The strands have been extended from right to left. So the sequence of a strand complementary to the original is found to be GGTGCA.

FIGURE 2.9 *(continued)*. *DNA Sequencing*

(b) Here the same procedure is done in four separate experiments, each with a different "labeled" dideoxy base. Each mix gives rise to distinctive "labeled" fragments, which form part of the DNA complementary to the original DNA strand. When these are run through different lanes of a gel, it is possible to see what kinds of bases terminate the shortest DNA strand, the second shortest, the third shortest, and so forth. So the sequence can be "read off" the gel. From the complementary sequence it is, of course, possible to deduce the original sequence, using the rules of base pairing.

ile X patients are unable to function outside institutionalized settings.

Not all mutant alleles show this simple pattern of differences. Investigations of Tay-Sachs disease reveal that there are several kinds of change in the genetic material that can cause the terrible degeneration. Cystic fibrosis turns out to be even more complicated. The sequencing of the gene disclosed a common mutation, found in just under 70 percent of CF carriers, that involves a three-base deletion from the normal sequence (the triplet corresponding to the 508th amino acid in the standard protein encoded by the gene is missing). The remaining cases have turned out to be heterogeneous, and over 200 further mutations at this locus are now known. Some combinations of these mutant alleles produce symptoms very like those caused by two copies of the predominant mutation. Some are highly deleterious. A few combinations are relatively benign.

All the diseases discussed so far are the result of changes at a single locus. One of the commonplaces of genetics is that the majority of traits of any organism—and human beings are no exception—result from the action of many genes and interactions between the developing organism and its environment. Nevertheless, disorders in some organ or tissue may be brought about because a mutant allele (or combination of mutant alleles) interferes with the normal process of development: For example, the retina, the central nervous system, or the digits may not form as they should. Facets of an organism that are brought about as a result of many processes may be disrupted in a wide variety of ways—plenty of things can go wrong in building eyes, brains, or limbs—so that what we pick out as "the same disease" may have dissimilar causes in different people. Restricting our attention solely to the genes, the disruption may be produced by a combination of alleles at one locus, by a single dominant mutation at a second locus, by a combination of mutant alleles at two quite different loci, by many combinations of alleles at three loci, and so on. More commonly, even combinations of alleles that enable some of their bearers to thrive may cause trouble for those who inhabit particular environments. An important moral is that some instances of "the same disease" may be the result of genetic changes at a single locus, as with the conditions just discussed, while others may come about through gene-environment interactions, or even because of features of the environment that are largely independent of the genotypic contributions.

These complexities are illustrated by a number of different cancers. Cancers occur because of misregulation of the mechanisms for cell division, so that a cell line goes into overdrive, producing an abnormally large number of descendant cells. Misregulation is caused through the activation of some genes and the loss (or modification) of others. Typically, the disease occurs only after many changes in the DNA ("multiple hits," as cancer researchers say). During the course of a human lifetime, there are many mitotic divisions (the ordinary cell divisions that go on all the time in our bodies), and at each division there is a small chance that some crucial piece of DNA will be altered. Environmental agents can trigger these changes, or they can activate a gene that would not normally be switched on. Genetic susceptibilities may also play important roles. Some people carry alleles that dispose their cells to tolerate copying errors in DNA replication. Sometimes this occurs only if their cells are exposed to particular environmental agents; in other cases, miscopying happens in a wide variety of environments. Other individuals may be missing a piece of DNA that plays an important role in suppressing tumors. If, say, six modifications of the DNA are needed for the tumor, these unfortunate people have taken the first step at conception, and only need five further "hits." In some individuals cancers result because of the action of environmental agents—such as radiation or substances that greatly increase the probabilities that their mitotic divisions will go awry. The "same" cancers occur in others because they bear alleles at several loci which collectively permit inexact copying at some critical areas of the genome. Yet others contract cancer because they carry from conception a nonfunctional tumor suppressor gene or genes whose protein products can activate an *oncogene* (a tumor-promoting gene).

Doctors have long recognized that some cancers (breast cancer, ovarian cancer, and colon cancer are well-known examples) run in families. Research in recent years has underscored the power of molecular genetics to identify genes implicated in some of these forms of cancer, and in particular, genes for colon cancer and early-onset breast cancer have been cloned, mapped, and sequenced. It is now possible to use genetic tests to determine if a particular person has one of the alleles that confer high risk for these types of cancer. Of course, many people—indeed, the overwhelming majority—who contract the cancers do not have the allele. Their cases are caused either through gene-environment interactions involving other alleles that confer susceptibility to the disease or through the insults of the environment. By the same token, presence of the allele only confers high risk, and some of those who carry it will be fortunate and never develop cancer.

As a final example, let us consider cholesterolemia, which plays an important role in some cases of heart disease. Failure to break down molecules of one form of cholesterol (LDL, low-density lipoprotein cholesterol, so-called bad cholesterol) leads to the buildup of cholesterol in the arteries, ultimately causing that clogging which, left untreated, culminates in strokes and heart attacks. The function of breaking down the cholesterol is carried out by the liver. People who carry a mutant allele for the LDL receptor gene have livers that do not perform the job efficiently and so suffer from hypercholesterolemia. Only about 10 percent of those affected with hypercholesterolemia carry a mutant LDL receptor allele, and although excess levels of LDL cholesterol increase the probability of heart attack or stroke, many people who have high levels of LDL cholesterol do not have cardiac problems. In this case, as in the examples of breast and colon cancer, genetic testing cannot deliver definite "yes or no" answers; the best that individuals tested can be given are probabilities.

Because of the virtuoso techniques of mapping, cloning, and sequencing, the genetic report card of 2020 is no idle fantasy. Biomedical researchers can seek out loci that affect human health, and they can determine the structure of the DNA that each of us carries at any of those loci. So my imagined parents can learn that their child certainly does not carry a number of very rare—but terrible—disorders, that the chance of some dreadful syndromes is not just minute but zero. For most of the ills that flesh is heir to, they will be told where things stand with respect to the normal risks: Here the chance is greater, there it is less. Is this information, the by-product of the molecular revolution, worth having? Will the hundreds of tests developed by probing particular regions of DNA prove liberating—or painful?

3

To Test or Not to Test?

Forewarned is sometimes forearmed. Knowing that a child carries an abnormal mutation may make it possible to act, even in ideal cases, to transform a life by substituting normal development for future suffering. All too often, however, there is no postdiagnostic magic. For those who receive grim news—learning, for example, that they carry the Huntington's allele (the over-long repeated sequence)—the only gains come in the ability to prepare for a truncated life. The stouthearted may find scraps of comfort in planning for the best under cruel constraint. Others less resilient will be crushed, unable to carry on with enterprises that might have brought satisfaction. For them, perhaps, ignorance is bliss.

As newspapers continue to report the discovery of genes implicated in human diseases, they announce enthusiastically that tests for susceptibility to those diseases will soon be possible. Journalists and their readers are excited by the prospect of relieving pain, incapacitation, and misery—perhaps so uplifted that they take it for granted that the tests will approximate those perfect examples in which diagnosis sets the stage for avoidance. So it is worth beginning by scrutinizing the assumption. Are there *any* ideal cases?

Enthusiasts often point to one shining example. In many parts of the affluent world, children are routinely tested at birth for phenylketonuria (PKU), a disease which if left untreated leads to severe mental retardation. Until very recently, the tests looked for high levels of the amino acid phenylalanine in a baby's blood, and although early versions of the tests could yield false positives, advances in our knowledge of the underlying genes now permit extremely accurate diagnosis. Children who carry two copies of a recessive allele are unable to produce an enzyme needed to metabolize phenylalanine with the consequence that a normal process that converts phenylalanine to another amino acid, tyrosine, does not occur in their cells. If they are reared on a normal diet, they come to have abnormally high levels of phenylalanine, and correspondingly low levels of tyrosine. These chemical imbalances in the brain produce the severe disruption of cognitive development. However, if the children are given a special diet (low in phenylalanine, high in tyrosine) from birth through adolescence, they develop almost normally. Children who do not have the PKU genotype are harmed by receiving the special diet, suffering comparable retardation to untreated PKU children. So here apparently there is a perfect test, one in which the population is divided into two groups and each receives just what is needed for normal (or near-normal) development.

Not quite. Taking a hard look at the history of PKU testing, the historian of biology Diane Paul has shown that the actual benefits are far less clear. As many doctors would acknowledge, the early days of PKU testing were fraught with mishaps. Because of the incidence of false positives, some children were assigned to the special diet and suffered irreversible damage; because nobody knew how long children needed to stay on the diet, many of them were returned to normal foods too soon, with serious effects on their development; because girls with the two mutant alleles had only reproduced rarely (if ever), nobody foresaw that pregnant women who had been liberated from the PKU diet would give birth to babies with catastrophic neurological defects. These might be acknowledged as the "growing pains" of the test, tragedies that happened on the way to a world in which similar tragedies can be completely prevented. For almost half a century, diabetics have lived longer, richer lives because their doctors have known how to prescribe and supervise insulin treatments. It is easy to forget that the early stages of insulin therapy were fraught with difficulties. (In fact, I owe my own existence to the strength of my mother's constitution and her ability to survive a period during which it was hard to tell an incipient sugar coma from insulin shock.) Does the analogy hold for PKU? Diane Paul has probed further, asking if, even today, we know just how much good—and how much harm—testing for PKU has done.

The PKU diet is both unpleasant and expensive. Especially in adolescence, it disrupts the social lives of those who must follow it and, consequently, both children and their parents have to exercise considerable discipline. Children on the diet have sometimes been placed in remedial classrooms (even though their academic performance is indistinguishable from that of their peers), and in many instances they feel themselves inferior to their contemporaries who can eat normally. How many children assigned to the diet actually manage to stick with it? How many families give up because it is too expensive? How many grown women with the PKU genotype who become pregnant know that they should return to the diet—or, if they know, how many have the strength of will and the economic resources to do so? These are questions whose answers we do not know, for the lives of those who have tested positive for PKU have never been systematically followed. In principle, it is possible for PKU testing to do enormous good for the very small number of people who carry the two mutant alleles. In practice, it is possible that testing has so far brought much less benefit, partly because of incomplete knowledge during the early years of PKU testing, partly because of a lack of a system of social and economic support for families who face the difficult task of integrating the special diet into their lives.

Even when medical research seems to have yielded an almost ideal test, even when the strictly medical problem seems solved (or almost solved), there are serious questions to be asked. What can we learn from the history of PKU testing? How can we build a social context in which genetic tests realize their potential to enhance human lives?

The examples discussed so far in this chapter illustrate some potential difficulties associated with genetic testing. One worry centers on the quality of the information a genetic test may offer. Barring processing errors, which are no more or less likely to affect genetic tests than other medical procedures, analysis of a person's DNA should tell us with perfect accuracy how the sequence of nucleotides he or she carries compares with normal and mutant alleles at the loci investigated. In diagnostic contexts, that is precisely what we want to know. Already aware that the patient is sick, we seek insight into potential genetic causes. (Indeed, not only do the advances of molecular genetics provide powerful new ways of diagnosing genetic causes of disease, but they sometimes increase our ability to recognize infectious agents.) When we hope to make predictions, however, matters are more complicated. The goal is to advise about the chances of a future phenotype, something straightforward to achieve—so long as we know the probabilities with which the bearers of particular genes acquire the worrying traits.

Unfortunately, we usually know something rather different but easily confused with the needed probabilities. Gene hunters proceed, quite reasonably, by concentrating on families in which those afflicted with a disease always (or virtually always) carry a genetic marker. Once they have captured, cloned, and sequenced their gene (or genes), researchers may find that everyone in their sample who has the disease carries the pertinent genotype. Further investigation might even show that all sufferers from the disease (including the many who were not originally sampled) have that genotype. Should we conclude that the genotype is the mark of doom? Not necessarily. The fact that the probability of having the genotype if you have the disease is 100 percent does not mean that the probability of acquiring the disease, if you have the genotype, is 100 percent—or even that it is significantly greater than zero. (After all, the probability of having a brother or sister if you have a twin is 100 percent, even though the probability of having a twin if you have a brother or sister is small.) People with that genotype might populate the world in large numbers, the overwhelming majority as healthy as you please, and only a small minority accounting for all cases of the disease.

Our tendency to confound probabilities is well illustrated by a recent episode in the history of behavioral genetics. About two decades ago investigation of the chromosomes of men in prison and in mental hospitals revealed an unexpectedly large percentage of men with an extra Y chromosome. So was born the idea of the criminal chromosome. Mothers who discovered through amniocentesis that the fetus they were carrying was an XYY male came to believe, on the basis of apparently rigorous science, that they would be likely to give birth to a future jailbird or maniac.

Subsequent research revealed that XYY males occur with significant frequency in the general population; indeed, that 96 percent of men with the extra Y chromosome lead normal, peaceful lives, never having problems with the police or suffering mental disturbances. So the probability that a fetus diagnosed as XYY is destined for the prison or the asylum is not high—it is only 4 percent. I suspect that nobody knows how many mothers were warned by their doctors that a fetus carried an extra Y chromosome, how many chose to terminate the pregnancy on those grounds, or how many would now change their minds.

Imagine that the results of amniocentesis and DNA analysis show that you or your spouse is carrying a male fetus with the Fragile X genotype: 400 copies of the CGG repeat are present. How likely is it that the child who would be born would be developmentally impaired? Because we know that boys and men with the syndrome have more than 300 copies of the repeat, it is natural to foresee some degree of mental retardation and the other emotional and behavioral problems typical of Fragile X. So far, however, the general population has not been systematically studied. It is possible, although researchers think it highly unlikely, that there are thousands of men with long repeats who are now leading seemingly normal lives, perhaps as stockbrokers, lawyers, or university professors. We do know that in families where one male manifests the syndrome, no males bearing the long repeat are phenotypically normal, a fact which provides some support for thinking it highly probable that a boy with the long CGG repeat will show the behavioral symptoms. Given the gravity of the decision whether or not to continue the pregnancy, however, wouldn't you want to be *sure* that the association between the extra-long repeat and mental retardation is not just a feature of particular families, a result of interactions among genes or between genes and a common environment? Wouldn't you want statistics about the general population?

To provide the information people need to make responsible decisions, genetic tests must be accompanied by reliable estimates of the chance that someone with the genetic condition will become afflicted with the disease. Those estimates can be deduced, using elementary laws of probability, from knowledge of the rate at which the genetic condition occurs in people who do *not* have the disease. Sometimes it is easy to see how to obtain this knowledge. Men who have the mental, emotional, and behavioral problems associated with Fragile X syndrome show the symptoms during adolescence, at the very latest; consequently, men in their twenties and older who do not manifest the symptoms can serve as an appropriate group for sampling, and doctors could simply take blood or a cheek swab from their male patients who fall into this class. (Doing so would not necessarily invade the privacy of patients, since samples could be marked anonymously as "asymptomatic.") But other diseases do pose problems. Discovering the rate at which bearers of a particular genotype contract colon cancer, for example, would require studying a very large group of people over a long period of time, for there is no age of which we can say, "If the disease is going to strike, then symptoms should be present now." Assembling the relevant statistics will always be time consuming and expensive, sometimes very time consuming and very expensive, but we should think of the investment as a necessary overhead for genetic tests, required if they are to inform responsible decisions.

In fact, if decisions are to be based on more than hunches or stereotypes, people should be given statistics relevant to their situations. Imagine a different scenario, one in which a couple learns that the fetus the woman carries has two copies of the common cystic fibrosis mutation. The doctor explains, "The probability that your child will have cystic fibrosis is one hundred percent," conjuring up the stereotype of a frail child dying in early adolescence. Although it represents one possibility, the stereotype is increasingly inaccurate: The median life expectancy is now twenty-nine, and many people with cystic fibrosis are surviving into their forties and fifties. The couple would be helped if they knew something about average life expectancies, about the chances of particularly severe and particularly benign forms of the disease. Even more, they would benefit from knowing the odds for children in the kind of environment they can expect to provide. One relevant factor (among many) is the fact that the disease has been diagnosed early. The parents would be quite misled if they were informed of probabilities based on averaging across disparate cases in most of which children suffered irreversible damage because diagnosis occurred too late.

Attitudes toward Down syndrome illustrate the dangers of thinking in terms of stereotypes or the very crudest kinds of probabilities. Two decades ago many doctors and prospective parents treated an extra chromosome 21 as if it were a signal for abortion, believing that the life of any child with Down syndrome must be highly restricted. Since then, dedicated parents, doctors, and social workers have effected large transformations in the lives of some Down syndrome children. Thanks to their efforts, when prospective parents are told that a fetus carries an extra twenty-first chromosome, they should no longer assume that a child born of the pregnancy is destined to a life of small accomplishment and low quality. Yet what exactly should they think? It would be wide-eyed romanticism to leap to the opposite conclusion, to think that given enough dedicated support, the child would add to the small number of admirable stories. Prospective parents need to know the full range of phenotypes associated with the condition, the full variety of functioning found in people with an extra twenty-first chromosome, and the probabilities associated with each in the best environment they are likely to be able to provide. Assuming the best, assuming the worst, and assuming the average are all mistaken. We would be better off in a world without lies or damned lies, but we cannot manage without statistics.

Assembling the proper informational context for a single genetic test is a significant task, requiring us to sample broadly from the general population and to differentiate environmental factors that are relevant to the expression of the disease. If the genetic report card of 2020 is to improve decisions, then it will need the backing of mountains of statistics, which will require considerable ingenuity and immense labor to collect. Without appropriate statistics, parents may leave the pediatrician's office believing that they have the power to promote their children's health, an illusion born of the convincing precision of meaningless figures.

Refinements in thinking about the likely impact of mutant alleles are all very well, but the ultimate point of genetic testing is to make a difference to the lives of those who are tested. Settling issues concerning the reliability of tests is only a preliminary to considering what people might be able to do if they received accurate and informative results. In some contexts, the knowledge is crucial to the actions of doctors or patients: A genetic diagnosis informs the doctor of how to treat the patient, a couple learn that they are at high risk for conceiving afflicted children and decide to adopt, expecting parents resolve to abort an affected fetus. (Of course, as we shall see later, each of these actions raises significant social and moral questions, but they show that sometimes there is something that *could* be done with the genetic information.) However, predicting the course of a person's life may or may not permit actions to guard against threats to future health. From the purely medical point of view, setting aside the social and economic factors that affect how the children grow, the test for PKU is a wonderful success story. Yet this rare condition owes its prominence in discussions of genetic testing to the fact that it is atypical. Many genetic disorders are more like Huntington's disease—those who test positive can, at present, do nothing to stave off future suffering or degeneration. Even when advice about possible changes in habits can be given, it is sometimes just vanilla advice. By identifying genetic conditions that put people at risk for certain types of cancer or for heart disease, doctors can tell their patients to exercise, to avoid fatty foods, and not to smoke, but these suggestions would have been available and appropriate whether or not the tests had been administered. If genetic testing performs any function in promoting health, it can be only through stiffening the resolve of those who are drawn to nicotine, red meat, and hours of television-induced stupor.

Present understanding of the causes of cancer makes clear why many recommendations offered to those who are genetically at high risk would be good for everybody. If cancer comes about through "multiple hits" (a number of changes in pertinent segments of DNA), it is in each person's interest to avoid the environmental causes of further "hits," whether or not one blow was already struck at conception. But besides trying to avoid exposure to carcinogens, people may undergo monitoring for incipient tumors, and thereby attempt to detect the disease at an early stage. Genetic tests for colon cancer promise to make a genuine difference. Routine inspections of the colon through colonoscopy are both intrusive and costly, and thus inappropriate for everyone in the population. However, because a program of annual examinations would detect growths before full-blown cancer has developed, those at high risk for colon cancer would benefit from monitoring. Tests for a genetic predisposition to colon cancer would identify the potential beneficiaries of periodic colonoscopies before the age at which they are recommended for everyone.

It is useful to compare tests for a predisposition to colon cancer with the envisaged test for an allele implicated in early-onset breast and ovarian cancer. What exactly can a woman do when she discovers that she bears an allele that makes her susceptible to breast cancer? Inspired by the example of colon cancer, we might think that she should perform self-examinations and begin having mammograms at an earlier age than women who lack the allele. Breast cancer is sufficiently common and the consequences sufficiently grave to warrant recommending self-examination to *all* women, whether or not they bear the allele. Setting aside questions about the possible side effects of mammograms, we must still acknowledge significant limits to the procedure's ability to detect small tumors, especially in younger women. As menopause approaches, breast tissue becomes less dense and more fatty, so that mammograms provide clearer pictures of potentially dangerous changes. But for most women in their twenties and thirties, mammograms are virtually useless. If a young woman with the unfortunate allele has the dense breasts typical of her age, the advice to increase the frequency of mammograms has little value. If she is unusual in having breasts that permit reliable analysis by mammogram, then she can promote her health by having frequent checks, whether or not she bears the allele. Women's decisions to have regular mammograms would better be based on the ability of mammograms to tell them something than on the findings of genetic tests.

All of this is true for the general population. Within those families with a history of breast and ovarian cancer, the test might be much more valuable: For some young women, a shadow would lift, while others might be prompted to the radical option of bilateral mastectomy (although, tragically, there are known instances in which this has not proven effective).

Plainly, the medical benefits of genetic tests have to be carefully evaluated. Biotechnological companies, having invested in developing new genetic tests, will be interested in ensuring that the tests are widely adopted. Doctors—especially the large majority who are not well versed in genetics, especially those in countries like the United States, where the threat of malpractice suits is omnipresent—may well take the line of least resistance and automatically recommend genetic testing, whether or not it serves their patients' interests. Tests may even become institutionalized in schools, so that children are routinely tested—and sometimes labeled—whether or not the knowledge proves to their benefit, whether or not they or their parents would have desired to learn the results. Equally unhappy would be a blanket skepticism in which genetic tests were viewed as expensive luxuries of no discernible value. The far-from-simple truth is that different tests can provide medical benefit, to different degrees, for different groups of people, but without guidance, doctors and patients alike will be caught between the pressures of an economic market and the angry (or bewildered) reaction of disappointed consumers. Tests must be evaluated on their merits, and the medical practice of the affluent democracies would best be served if there were independent bodies that could issue clear assessments of the medical uses to which tests can be put.

Such assessments should not assume that what can be done in principle will be representative of what is done in practice. One of the morals of the history of testing for PKU is that the actual benefits achieved by introducing a test may fall short of those that might ideally be attained. As the reviewing body ponders a new test, its members must ask themselves hard questions: Not only, "Is there a procedure that would bring medical benefit to those diagnosed?," but, "Could people in our society afford it?," "Would they be able to continue with it?," "Would they be given the continuing medical counsel they will need?" Realistic assessment of the answers may lead the reviewers to a more somber evaluation. Or maybe those answers will inspire the citizenry to provide funds to change the system of social support, to give subsidies to those who need expensive treatments, to resist the pressures that make it hard to undergo the treatment—exemplified in the costs and social disruption of the PKU diet—to offer continued help and guidance for those who test positive.

• • •

Genetic tests to identify those at risk of developing a disease might offer a quite different type of benefit. Even in those cases in which nothing can be done to avoid the disease, greater knowledge can bring greater autonomy. Well-informed people are better able to plan their lives, to recognize their potential limitations, and to shape for themselves futures that best achieve what is most important to them. The possibility of nonmedical benefit from genetic testing is clearest when the test issues a nonprobabilistic verdict. Learning that you carry the Huntington's long repeat is a truly terrible discovery, even if you have known since early adolescence that you are at risk for the disease. Still, it might sometimes be better to know than not to know, and to use that knowledge to plan a career, a family, a life that accommodates the remorseless fact.

Since the arrival of tests for Huntington's disease, it has become apparent that many fewer people have opted for testing than the relatively large proportion of the North American population (around 80 percent) who predicted that they would want to be tested. Concrete opportunities have a way of focusing attention on the disadvantages of an offer that was attractive at a distance. Some people know that they would be crushed by bad news. Others, possibly many more, merely doubt that they can predict their own reactions, and realizing that their lives might be derailed, forego the chance to plan in knowledge of the worst (or, more luckily, the best). In the abstract, we are too quick to swear allegiance to the ethos of bravely looking truth in the face. Although some people may be confident that they would not blink, it is quite reasonable for many to hesitate—and to decline to take the test.

Because human self-knowledge is limited, any widespread introduction of genetic tests will need a system of support, so people may be guided to a better understanding of their own needs. Doctors who are intimately involved in genetic testing for the most devastating diseases and disabilities know all too well the power of information to disrupt lives: After learning that she carries on one of her X chromosomes the translocation that is responsible for her brother's state of hapless infancy, a young woman breaks off her engagement and foreswears further prospects of marriage; she feels that her life is ruined. Even if there is ample pretest counseling from people who are sensitive to the possibilities of grave psychological damage, some cases like this are bound to occur. Without a well-developed background counseling program, however, the potential for administering tests to people who cannot cope with the bad news is enormous.

As it becomes possible to deliver a wider variety of devastating verdicts, the explanations and conversations offered by genetic counselors will have to be much more detailed. It is far too easy to conjure up the image of an ideally rational person, fully self-aware, who has thought through the possible outcomes in advance and is ready to make maximal use of the information presented. Real counseling sessions rarely involve such individuals. Very frequently, those counseled are not sure why the tests have been recommended, or are confused about what the tests show, or are unable to understand basic points about probabilities and risks. Studies of genetic counseling sessions show that much of the time counselors are not aware of the problems their clients want to discuss and, equally, the clients find the explanations given incomprehensible.

Communication breaks down even when the people being counseled are affluent and well educated. But tests and counseling support should not be designed to fit a stereotypical client. We should not confine our thinking to the wealthy, to people who have resources that will suffice to translate information into action, to those for whom the uncertainty at which the genetic test is directed is the focal problem in an otherwise serene life. People with these characteristics may now be overrepresented in genetic counseling sessions, but they are hardly representative of the population, even in the wealthiest societies. If only because all citizens pay for the research, genetic tests should serve the needs of all; it would be quite wrong to devise a system that would serve only the most privileged.

Members of ethnic groups most at risk for stigmatization and other adverse effects through the dissemination of genetic information are also least likely, as things now stand, to benefit from genetic testing. In some countries, most evidently the United States, their access to the medical system is more restricted, and they may lack resources to follow the recommendations that stem from genetic diagnoses. Medicine may be too expensive, lifestyle changes impossible within the other constraints on their lives. However strong their ability to take stock of bad news, they may be so assailed by other forces that the genetic test is essentially irrelevant. Women who are victimized by spousal abuse, who are in danger of eviction, or who struggle to bring up children whose fathers have long deserted, face problems that dwarf the threat of diseases that may strike a decade hence. Unless the patients' background situations are taken into account, genetic testing cannot be expected to play a positive role in their lives.

All too often patients have forfeited their autonomy before the test begins. For all the efforts of genetic counselors, clear communication of the consequences may prove impossible. Many people find themselves at a loss in medical situations and are pushed in directions they do not understand and would not have chosen for themselves. Genetics is hard to grasp, especially when it involves understanding probabilities, and even the well educated sometimes leave counseling sessions mystified and uncertain about just what they have been told. As genetic tests proliferate, it will be especially important to develop effective ways of communicating important news to people whose education is rudimentary, who are unfamiliar with medical terminology and daunted by medical settings, who do not conceive the possibility of defying doctor's orders.

At present, accepted guidelines urge genetic counselors to be "nondirective," honoring the attractive idea that counselors should not impose their own values, but "stick to the facts," leaving those they counsel to make decisions that are genuinely their own. Although this may be a splendid ideal, the transition to societies in which genetic testing is more prevalent will almost certainly multiply the number of situations in which "respecting the autonomy of the counseled" would be a bitter joke. Counselors will sometimes be the last and only possibility for articulating a course of action for people whose awareness of their own predicaments and options is, at best, partial. They will have to learn how and when to elicit the decision that most accords with their client's interests and values, neither inserting their own priorities nor remaining so remote that the advice they give is useless.

Critics of genetic testing often conclude from the fact that testing does not help (and might even harm) large segments of the population that the development of further tests should be resisted. Champions of genetic tests, recognizing the value that tests can bring to (stereotypical) fully informed, autonomous individuals, can make the plausible rejoinder that molecular medicine cannot be held responsible for failures in systems of social support. Genetic tests can provide benefits: preventing the birth of babies who would suffer the degeneration and early death of Tay-Sachs and Canavan's disease, enabling doctors to improve the prospects of some patients with high cholesterol, even allowing the stoical future victims of Huntington's disease to organize better lives for themselves. Developing new genetic tests is a good thing, and if the delivery of the tests proves problematic because of current social inequities, then the remedy is not to halt the progress of molecular biology but to attend to the social problems.

Although the zeal of the critics may be misdirected, the simple reaffirmation that molecular medicine can help some people prevent future disease or avoid future tragedies is too complacent. As we shall see in more detail later, the broad program of which genetic testing is a part—the immediately achievable part—does promise a significant shift to a form of preventive medicine that would be well worth the efforts of attaining it. Without attention to the social surroundings in which molecular medicine is practiced, however, that outcome would simply magnify inequities that are already present.

Devising better ways to focus the aims of medicine and to frame the social context of medical practice is a critical task in allowing molecular medicine to fulfill its promise.

A particular way of thinking has so far dominated the discussion. I have imagined the ideal future of genetic testing as if our descendants would have the opportunity of visiting a vast supermarket in which genetic tests were offered. If the supermarket is properly run, the wares will come with detailed manuals for their use (background statistics) and advertising of impeccable accuracy (independent assessments of medical benefits); the staff (well-trained counselors) will guide the customers to those products that best suit their needs; nobody will be forced to buy anything.

Should there be limits on consumer freedom? Are there some tests that everybody should be required to take? Are there tests that ought to be forbidden? Moved by respect for individual liberty, we might give automatic negative answers. However beneficial a test might be for someone, that person cannot legitimately be compelled to take it. By the same token, nobody should be prevented from taking a genetic test, even if that test is directed at something entirely frivolous. Only when it becomes clear that the person is incompetent are we justified in intervening—and making a decision against one's own best interests (or others' perceptions of those interests) is not an automatic sign of incompetence.

But liberty is limited, as John Stuart Mill argued eloquently, by the effects of one individual's actions on others. Even if we should not compel someone to take a test that could provide potentially life-saving information, perhaps genetic tests can justifiably be required when the well-being of third parties is involved. Imagine, for example, a single mother with very young children, who is known to be at high risk for a disease that is preventable by early detection. Her premature death would have emotional consequences for the children left behind, and it would be reasonable to try to make her vividly aware of these consequences. Yet if in the end she shows clear understanding of the points that have been made, admits to their force, and is deeply moved by the envisaged plight of the children but remains convinced of the overriding wrongness of breaking religious principles which she devoutly accepts, seemingly the state has neither the practical ability nor the moral right to compel her to take steps to avert the disease. Following their own sense of what is important and valuable, parents sometimes expose themselves to dangers that have the potential to affect their children. Sometimes these decisions are irresponsible, as when medicines are not taken regularly; sometimes, they are constitutive of the person's life, as when mountaineers or undercover narcotics agents go about their business. Even if the genetic test were administered, it might well be impossible to ensure that its deliverances were correctly followed. Moreover, in the end, a state must acknowledge a citizen's right to engage in risky behavior when the activity is somehow central to that citizen's life conception. If disaster comes, then we must clear up the mess as best we can, ensuring that the suffering of innocents who are affected—such as the children in the imaginary example—is minimized.

Sometimes, however, the harms inflicted cannot be avoided through later interventions by others. Besides the practical difficulty of compelling compliance with health recommendations, what allows us to respect the right of citizens to self-determination is, I suggest, the possibility that the sufferings of the innocent can be significantly reduced. The state can ensure (if it has the will) that the orphaned children receive care and nurture. By contrast, if the child has PKU and is not treated, severe retardation is inevitable (and, as noted, if the child does not have PKU and is given the special diet, then the consequences will be equally grave). However sincere the mother's religious convictions, she cannot be allowed to wreck the life of a child, and in the extreme case—when the result of the test for PKU is positive—it would be justifiable to remove the child to ensure that the special diet is provided. (But we must also ask if "rescuing" the child would be hypocritical in societies that do not provide subsidies for families to buy the PKU diet.)

How far does the right to intervention extend? Can we justifiably require prenatal testing and termination of pregnancies that would culminate in the birth of a child whose life would be brief and agonizing? With the best intentions, we are easily led down a road that our predecessors have traveled: We start by trying to avoid human suffering and then inch our way to full-blown eugenics. Full discussion of the complex issues surrounding eugenics will occupy us later. For the moment, I shall simply explore whether allowing for the overriding of individual freedom in extreme cases—to assure proper treatment for children with PKU, for example—sets us on a course to far more dubious invasions.

Required genetic testing for PKU appears defensible because we rank the restriction of parental freedom as a lesser harm than causing irreversible damage to the child. The same rationale would require prenatal testing and treatment of a fetus in order to prevent tremendous damage, provided that the woman also had the opportunity to terminate the pregnancy. Geneticists already know the basis of a number of other syndromes, all of which bring severe mental retardation and behavioral problems. In Lesch-Nyhan syndrome, for example, an allele on their single X chromosome causes boys to suffer mental retardation and extreme physical pains of the kind associated with gout. Yet perhaps the most moving and disturbing feature of the condition is an apparently irresistible urge to self-mutilation—the boys chew their lips and the tips of their fingers until they are raw and bleeding. At present, doctors can relieve some of the gouty symptoms, but they are unable to prevent the mental retardation and can only block the compulsive mutilation by applying bandages to hands and lips. We can envisage future molecular geneticists identifying the presence of the Lesch-Nyhan allele prior to the point in development at which that allele plays its most important role and devising medical procedures to prevent the syndrome.

Imagine a woman who is at significant risk for bearing a child with Lesch-Nyhan and who wishes both to continue the pregnancy to term and not to have the test (or the preventive treatment). We can acknowledge her right to make her own reproductive decisions by ensuring that she has the chance of terminating the pregnancy at this stage. Although the ability to make free and informed reproductive decisions may be far less widespread than is commonly supposed, the woman has a right to make her own decision in light of the best available information, and the state has a duty to create conditions under which this can occur. Once she has decided of her own volition to continue the pregnancy, she has made a commitment to the well-being of the child who will develop from the fetus, and actions that would cause irreversible damage can be restrained. Genetic testing may be required, even though she refuses her permission, for exactly the same reasons as in the case of PKU.

The case just described is almost purely hypothetical, not only in its optimism about a future treatment for Lesch-Nyhan syndrome but also in the manufactured conflict between the woman and the fetus. The tendency, in many discussions of moral issues surrounding pregnancy, to oppose the interests of women to those of the fetus quite overlooks the fact that almost all women who risk harming the children who will develop do not want to cause such harm. Sometimes their actions arise from ignorance or from the stronger pressures that act upon them. Only very occasionally are they in the grip of religious doctrines that urge them to allow irreversible damage to a child who might have been born healthy. Interventions to help the fetus should typically be thought of as ways to help pregnant women who have chosen to let the pregnancy proceed to term achieve what almost all of them want—healthy babies—within the constraints of situations that are often extraordinarily difficult.

Sometimes, unfortunately, the pressures on women make it psychologically hard, perhaps impossible, to reach decisions that are in the best interests of the fetuses they carry. Poor women who resist testing for AIDS would surely prefer, other things being equal, to save their babies from suffering and early death. Yet other things are not equal, and terrified by the prospect of being abandoned or becoming outcasts, they avoid the test. Compelling them to be tested and to take AZT may be the right choice as a stopgap measure, since it may be a choice that saves a child, but it is only part of a complete response to the problem. In attempting to rescue a baby from AIDS, we should not be blind to the predicament of the mother, and we should work to remove the causes of her justified fears. (We might also ask ourselves whether heroic efforts at prenatal rescue are in harmony with toleration of the grim conditions in which many children—including, very probably, many of those to be saved from AIDS—grow up.)

The hypothetical case does show clearly that the rationale for requiring PKU testing does not extend to mandatory prenatal testing for conditions about which nothing can be done except to terminate the pregnancy. In the real case of PKU, and the imaginary example of Lesch-Nyhan, there is an identifiable individual to whom irreversible damage would be done in the absence of active intervention. By acting, we change the world to make a child's life far better than it would otherwise have been. When the only option is to terminate a pregnancy, there is no such identifiable person whose lot becomes better. Possibly, by intervening, we would make the world happier than it would otherwise have been, for a world in which the Canavan's carrier dies *in utero* may be a world with less anguish than one in which the degenerative disease takes its horrible course. But we have not brought any benefit or relief to the bearer of the foreshortened life. The rationale for requiring infants to be tested for PKU is based on commitment to a person who would be harmed by not intervening, a person who cannot express his or her own concerns. Prenatal testing where nothing can be done except to abort affected fetuses does not fall within the scope of that commitment. We can consistently support mandatory PKU testing for newborn infants while resisting compulsory prenatal testing for irremediable birth defects.

Just as the rationale for requiring tests is grounded in the idea of protecting third parties, so too we can only be justified in forbidding genetic tests if those tests would be used to cause third parties harm.

Despite the wide range of opinions about when abortion is permitted, very many people probably agree that certain kinds of genetic conditions do not provide an appropriate basis. Couples who tested their fetuses for the presence of blue eyes or curly hair and who decided to abort otherwise healthy fetuses when an alternative genotype was present would, to say the very least, have a distorted conception of value. Should states therefore limit the liberty of couples to make reproductive decisions? Not necessarily. Freedom of reproductive choice can be defended provided that the fetus is not taken to be a person with rights and interests that the state has a duty to protect. For the moment we can accept the idea of the testing supermarket, open to free choice with few restrictions. Later, we shall have to come to terms with people who think that the genetic supermarket will raise the incidence of abortion, thus causing murder of the innocent, and who conclude that morally responsible people should shut it down.

One last aspect of the question deserves note. Disability activists have been vocal in opposing widespread practices of prenatal testing, fearing that a program to lower the incidence of genetic disabilities will worsen the plight of those who are born with the conditions, both through the withdrawal of support and through the loss of respect. Although their concerns are readily comprehensible, existing examples of testing programs show that encouraging attempts to reduce the incidence of a genetic disease is compatible with continuing to respect those born with the disease and providing support for their distinctive needs—indeed with improving their lot. In Cyprus, the Greek Orthodox Church has administered a successful program designed to reduce the incidence of thalassemia (a disease involving malfunctioning hemoglobin, which produces effects similar to those of sickle-cell anemia). As the incidence of thalassemia has diminished, help for the afflicted has increased: Because there is now less demand for blood transfusions and other treatments, the lives of thalassemia sufferers are now better than they were. If affluent societies build and stock the genetic supermarket, allowing almost unrestricted use of genetic tests in reproductive decision making, even perhaps encouraging the prevention of lives burdened with genetic disease, it will be important to honor the principle that those who are born be given equal respect and the distinctive support they require.

Genetic testing can easily be botched. If we are in haste to produce the "genetic report card" of 2020, neglecting the statistical information that will make the figures meaningful, foregoing independent assessments that would guide patients and doctors to tests that could provide medical benefits, ignoring the need to provide genetic counseling on a far broader scale and in response to the predicaments of all citizens, then we may produce a monster that causes more disruption than improvement in the lives of our descendants. The handful of examples we can presently actualize points to the need for substantive changes in providing access to medical resources and in altering social arrangements that would cause havoc once genetic testing becomes widespread. Tests cannot merely be thrust into existing social milieus, their fortunes fixed by prevailing social and economic conditions. It would be tragic if we were to give up the real benefits that genetic testing can bring—but equally tragic if blind attempts to reap the obvious dividends were to go awry.

Those who hesitate at the prospect of the many genetic tests we can already envisage, and the hundreds, even thousands, that the next decades will make possible, are not Luddites, unmindful of the real benefits that molecular diagnoses might unfold. Cautioned by history, they worry that affluent societies will not commit the resources needed to make genetic testing work for all citizens. No intricate analyses are required to expose the ways in which molecular diagnostics could become an instrument for good, but the necessary programs will not be cheap. Do our societies have the will to make the investment?

4

The Road to Health?

Shortly after beginning his journey, John Bunyan's pilgrim, Christian, is joined by one of his neighbors, Pliable. Pliable seems enthusiastic about the enterprise until the pair fall into the Slough of Despond, and this early setback is quite enough to quench his ardor. Pulling himself out of the muck, he sets off home. Christian, by contrast, with a clearer understanding of the glorious goal, struggles through the mire, goes on to face many other reversals, and ultimately wins his heavenly reward.

Myopic pilgrims are unlikely to go far. Perhaps in concentrating on the short-term difficulties of genetic testing—the problems posed by situations in which we can predict that disease will strike but are helpless to avert or alleviate it—we are like Pliable in the Slough of Despond. We do not look beyond our current discomforts to see the more distant rewards. After all, our visionary comrades may remind us, although the ability to test for genetic disease is the first, most easily attainable application of the new molecular biology, the eventual hope is to be able to intervene, to treat, to cure. So, as we learn more about the genes that are implicated in diseases and disabilities, we shall find ways to bring relief. The road leads on beyond the dismal place in which we can test broadly but offer only a handful of therapies. Instead of accepting our present limitations as fixed, we should take heart from the promise of molecular human genetics.

Today the Human Genome Project, an international effort most heavily concentrated in North America, Britain, Continental Europe, and Japan, stands as the most visible symbol of that promise. The official goals of the project are to produce complete genetic and physical maps of the human genome (the totality of the genetic material in a typical human cell), and of the genomes of other organisms, to develop sequencing technology (better ways of finding the order of A's, C's, G's, and T's in a region of DNA), and ultimately to identify the full sequence of the human genome. This is to be the next stage in the transformation of medicine. Armed with knowledge of the full three billion base pairs, recognizing all our genes, we shall be able to understand how our bodies work and how they sometimes fail us. Eventually, champions of the project predict, we shall be able to intervene on a grand scale to prevent or to ameliorate human suffering. Another image of a great journey recurs in advertisements for contemporary biomedical research. No less a figure than Walter Gilbert, Nobel Laureate and coinventor of sequencing technology, has seen genome sequencing not as tedious drudgery but knightly enterprise: The sequence of the human genome is the "Grail."

Promises, promises. How exactly is human health to be advanced by this intricate molecular knowledge? What precisely is the significance of a three-billion-character sequence of A's, C's, G's, and T's, spread, perhaps, across a thousand bulky volumes or condensed in a computer memory? Since the first proposals were made in the 1980s, critics of the Human Genome Project have suggested that the claims about the future of molecular medicine are overblown. They concede that complete sequence information might have some limited value. Were an angel to arrive bearing tablets inscribed with a full human sequence, the gift should not be refused. But critics can think of better gifts that angels might bring, and they distrust the promise that this particular revelation will revolutionize our knowledge of ourselves.

When talking informally, virtually all those involved in human molecular genetics would admit that matters are far more complex than the enthusiastic slogans make them appear. They would surely resist the accusation that the Human Genome Project is all empty promise, but the complications of the work they do, day by day, make it hard to predict just what the ultimate benefits will be or how recalcitrant are the obstacles they must overcome. They journey in hope, not knowing for sure that they are playing Christian to the naysayers' Pliable, whether they are Arthurian knights or Don Quixote.

To achieve a realistic picture of likely futures, we need once more to step back, to look not just at the details of what molecular genetics has so far achieved, but at where it is currently going.

A jug of wine and a loaf of bread, two thirds of the preconditions for paradise, come into being partly through the activities of a single-celled organism, the yeast *Saccharomyces cerevisiae*. The same humble organism helps to build paradise for the molecular geneticist. Thanks to a fast growth rate and susceptibility to easy experimental manipulation, the classical genetics of yeast is well developed, setting the stage for today's biologists to apply their new techniques. They are confident of knowing before the end of the century the detailed sequence of the nucleotides in an entire yeast genome—two chromosomes have already been sequenced—and they look forward eagerly to the day when all the genes and their functions have been revealed. That day will bring a new level of biological understanding, as we begin to fulfill the dream of disclosing the secret workings of life. Among the mechanisms brought to light will be many that operate in all *eukaryotic cells* (cells that contain a nucleus), including the cells that make up your body and mine.

Yeast is not alone in the molecular geneticist's pantheon. Another place of honor is occupied by the nematode worm, *Caenorhabditis elegans*. Adult worms, about 1 mm long, are either hermaphrodites or males. Hermaphrodites have 959 somatic cells, and painstaking scientific effort has constructed a complete *fate map*, a depiction of exactly how successive cell divisions beginning from a single fertilized egg give rise to each one of those somatic cells. If we only knew the molecular structures of all the genes, if we could identify their protein products, then we could begin fathoming the details of the processes through which nematodes develop, some of them processes common to other multicellular organisms. But we are on the verge of identifying the entire sequence of the nematode genome—over 2 megabases (2 million bases or 2 Mb) of contiguous sequence from chromosome III is already known, and the international group collaborating on the project expects to sequence the entire genome within the decade. The researchers believe that the mysteries of nematode development lie just beyond their view.

Fruit flies have been at the center of genetics ever since 1910, when Thomas Morgan assembled milk bottles, rotten bananas, and some exceptional undergraduates in the famous "fly room" at Columbia University. In the last decades, *Drosophila melanogaster* has assumed new importance, as the ingenious experiments pioneered by Christiane Nüsslein-Volhard and her colleagues have exposed the basic pattern of the fly larva, revealing how, at a very early stage, the embryo is divided into parts that will grow into the various segments of an adult fly. Flies tell their heads from their tails thanks to a protein inserted into the egg by the mother, and the relative sizes of body parts—head, thorax, abdomen, and tail—are determined by interactions among various proteins, some originating with the mother, others transcribed and translated from the fly's own genes. The segmentation pattern is fixed by further genes, genes that are discovered through mutations that cause some segments to be skipped, producing malformed embryos that typically die very young. Piecemeal analysis of the protein products of these genes has produced a detailed story of basic processes in early larval development. Armed with a complete sequence for the *Drosophila* genome and a catalog of all the fly's genes, many molecular geneticists hope for still more, for understanding patterns of development that are less rigid, more open, than those of the nematode worm *C. elegans*. They expect to know the intricacies of how flies are made.

Enough to quicken the pulses of specialists, perhaps—but weren't we promised more? The jug of wine and loaf of bread are less interesting, by far, than thou. Can we expect to obtain detailed accounts of human functioning comparable to those envisaged and partially obtained for yeast and worms and flies? Can we even expect to sequence the entire human genome?

The successes of contemporary genomic sequencing inspire confidence, but yeast and the worm may mislead us. Their genomes are small (15 Mb for yeast, about 100 Mb for the nematode). Scale matters. Indeed, scale matters *enormously*. Highly talented, imaginative scientists are needed to direct the coordinated parts of a massive sequencing effort. Even with so great an investment of intellectual resources, the goal may not be attained. However, before we consider whether it is feasible to deliver the sequence of the three billion base pairs of the human genome, we should obtain a clearer view of the benefits that are supposed to ensue. If there were a sequence-bearing angel, what use could we make of the gift?

Enthusiasts sometimes envisage a royal road for twenty-first-century biology and medicine. Given the full human sequence, the first task will be to catalog all the genes. Equipped with a list of genes, we shall then use the genetic code to deduce the sequence of amino acids that compose the protein whose formation each gene directs. At the next stage, we shall need to work out the three-dimensional structures and the functions of the proteins. We shall then be in a position to analyze the interactions among proteins and nucleic acids in cells, to uncover the fine details of physiological processes, and to trace the exquisite molecular choreography that leads from a zygote to an adult human being. Knowledge, we may hope, will be power, and our understanding of the molecular bases of physiology and development will enable our descendants to intervene to prevent or to ameliorate disease.

Other visions are less ambitious, envisaging a future in which the pace of contemporary molecular researches is greatly accelerated, not because a picture of our workings is systematically built up, but because piecemeal projects can proceed more rapidly once the sequences of candidate genes are available. Months of work now undertaken to locate and to analyze molecules of interest for particular projects may be compressed into a few hours. So through the next century our understanding of bits and pieces of the human bodily machinery will increase, probably at an ever faster pace, and that understanding will be translated into medical procedures that bring relief to people who would otherwise have suffered.

Yet at first sight, it is not entirely obvious that sequencing the human genome will take us where we want to go. Recall that genes are sparsely distributed in the human genome (and in mammalian genomes generally); they are islands in a vast sea of DNA, much of it highly repetitive and apparently without function. Current estimates suggest that we have between 30,000 and 100,000 genes. If these genes, plus their flanking regulatory regions, average about 3 kilobases (3,000 bases or 3 Kb), then they amount to at most 300 Mb in a genome of 3,000 Mb, or about 10 percent. Why is it important to sequence the rest—especially when tracking through the repetitive wastelands is so hard? Despite the difficulties of large-scale sequencing, many researchers believe the great trudge to which they are committed serves an important goal: They want to identify *all* the genes. Patience counsels them to sort through the dross so that they can find every piece of genetic gold.

How exactly is the winnowing to be done? How can we pick out the genes among all the A's, C's, G's, and T's? Occasionally, as in the soil amoeba *Dictyostelium discoideum*, changes in the ratio of C's (and G's) to A's (and T's) reveal differences between gene-rich regions and the intergenic wastes. For most organisms, however, including ourselves, there are no canonical sequences that mark the beginnings and endings of all and only genes (or, even better, the beginnings and ends of exons, those parts of the DNA represented in the protein products), as if the sequence, "the ultimate map," was dotted with X's marking treasure.

Although the problems are hard, they are not insuperable. Two main techniques are available for finding genes from sequence data in organisms, like us, that lack helpful signals. One is to search a region of sequence for *open reading frames*, strings of bases that continue a long way without a stop codon: In nonfunctional stretches of DNA—areas that natural selection does not care about—stop codons can easily accumulate; in the functional segments—the genes—the intrusion of premature stop codons is likely to make a difference to the organism's chances of surviving and reproducing; hence long sequences without stop codons are good candidates for genes. In each chunk of DNA, six reading frames are available. There are two strands, which are potentially capable of being transcribed and translated in opposite directions; on each strand we can envisage starting at the first base, the second base, or at the third base. Because triplets of nucleotides code for amino acids, starting at the fourth base is equivalent to beginning with the first, starting at the fifth tantamount to beginning at the second, and so on. Out of the sixty-four triplets (codons), three are stop codons. On average, therefore, we would expect a stop codon to show up about every twenty codons (every sixty bases). Drawing on all these facts, computers can be set to work, analyzing the sequence data to direct attention to areas in which there are substantially longer open reading frames.

Once an open reading frame has been disclosed, there are various techniques for determining whether or not it contains a functional gene. In organisms that lend themselves to experimental manipulation, a straightforward approach is to search for phenotypic changes that occur when the sequence is modified. But looking for open reading frames is only one way to hunt genes in the sequence wildernesses. Another, currently more popular, is to use the knowledge we have to gain more.

Sequences of thousands of genes in the intestinal bacterium *Escherichia coli (E. coli)*, in yeast, in nematodes, in fruit flies, in mice, and in our own species are available in data banks. Some of the data comes from positional cloning (the map-directed strategy that has proven so effective in finding human disease genes), some from searches for open reading frames, some from using other forms of analysis, especially on organisms that allow experimental tinkering. Employing algorithms that seek similarities in sequence pattern (and thus indicate *homologies*, sharing of sequence through common evolutionary history), computer-literate biologists can explore regions of DNA and identify other likely genes. The more we find, the more patterns we can seek.

We can even find out how well we are doing. There are some regions of some genomes—in *E. coli*, yeast, and *Dictyostelium*, for example—for which a fairly complete catalog of genes is already available. Suppose that we take the sequence data for one such region and a modest database of gene sequences. We can then set the computer to run the algorithm, identify new genes, and so expand the database. Running the algorithm on the new larger database, we can then discover additional genes and iterate the procedure until no new "genes" are added. At this point, we can take stock, checking the extent to which we have found all the genes and only the genes. We can then refine our algorithms accordingly.

Gene hunting by way of homologies takes advantage of a fundamental feature of the evolutionary process. Organisms are cobbled together out of many of the same parts, putting the bits and pieces to different uses in different assortments. In recent years genome scientists have discovered such "conserved sequences" in flies and mice, yeast and humans. One striking discovery about human cancers was achieved through detailed knowledge about a DNA repair gene in yeast, and a gene implicated in obesity in mice is very similar to a human gene. The history of life plays all kinds of variations on a relatively small number of themes. New genes, DNA sequences that code for proteins not previously produced in living cells, do not appear by some magical concatenation of bases in the genome. They are typically built from copies of old genes that have been modified, expanded, truncated, or recombined. So we should not be surprised by our kinship with the brewer's friend: There was no need for our cells to reinvent the machinery required for the basic intracellular processes. All eukaryotes—yeast as well as humans—take over opportunistically the breakthroughs of their remote, common evolutionary past.

As a result, the bootstrapping procedure just described is biologically faithful, and its apparent success is not a lucky accident but something we might have anticipated. Of course, there may be "orphan" genes, genes that we shall never detect because of the characteristics of the gene sequences in our initial sample or because they involve strategies for ringing changes on old themes that are unrepresented in our algorithms. Yet as more complete information about the simpler genomes—*E. coli*, yeast, nematodes—becomes available, that information can be used to fine-tune our search procedures. To predict complete success in finding all the human genes would surely be foolhardy. But we may hope to come close.

How can we go further? Once we know the sequence of a gene, and its division into exons and introns (thus separating the parts that do and do not contribute to the finished protein), deriving the ordered sequence of the amino acids in the corresponding protein is automatic. Finding the three-dimensional structure of the protein is, to understate, far harder. Although crystallographic investigations enable the structures of some proteins to be identified, the general problem of predicting how proteins fold into their three-dimensional shapes remains unsolved. Fortunately, we can build on existing knowledge, just as in the enterprise of gene hunting. Using algorithms that recognize significant sharing of amino acid sequences, researchers are able to assimilate many "new" proteins to molecules whose structures have already been identified by the laborious techniques of X-ray crystallography, often to molecules whose functions are already known. Lucky investigators see that a fresh piece of sequence contains a coding region and, virtually simultaneously, specify the amino acid sequence, the three-dimensional structure, and the function of the protein. Evolutionary conservatism is sometimes kind. Certainly there are limits to our power to pull ourselves up by our bootstraps, but at present those limits seem relatively broad. Careful investigation of the genomes of nonhuman organisms should help to specify the structures and functions of a large number of proteins, and we can use this knowledge to understand a large number of *human* proteins.

According to a famous, probably apocryphal story, the great British biologist J.B.S. Haldane was seated at dinner next to an Anglican bishop who injudiciously inquired what Haldane's evolutionary studies had taught him about the mind of the Creator. The answer was surely disconcerting: "He has an inordinate fondness for beetles." A genomic scientist in Haldane's place would recall the vast number of variant molecules that cells employ to add phosphate groups to proteins—and reply with equal aptness: "He has an inordinate fondness for protein kinases."

In a narrow range of cases, a gene's causal contribution to cell, tissue, organ, even organism can be read off the limited biochemical function of its protein. Establishing that a gene codes for a DNA repair enzyme educates our guesses about the differences the gene makes to the phenotype at various levels. Organisms with a mutant allele will tend to contain cell lineages in which there are "mistakes" in the DNA sequence. Gross blunders may be lethal. Lesser errors are likely to show up in the misregulation of cell growth and division, most obviously in the formation of tumors. Hence the recognition of kinship between a human DNA sequence and a repair enzyme in yeast has led researchers to suspect—rightly—that they have found a human cancer gene.

If we were permitted to make directed changes in the DNA of human zygotes and to let the zygotes develop in carefully controlled environments, we could explore the effects of mutations in more systematic fashion by searching for phenotypic differences. Such so-called knockout experiments would be grossly immoral on members of our own species, yet we can do almost as well by exploiting our evolutionary relatives. Higher primates, the organisms closest to us, would offer the firmest basis for analogical inferences about the roles our own genes play, but ethical qualms aside, the length of time between their generations and the complexity of the gene-environment interactions in their development make them less than ideal. We need an animal close enough to ourselves to help us pinpoint the phenotypic effects of a large number of loci, an animal that breeds rapidly, that can be reared in a number of relatively simple environments, and whose genetics is relatively well understood. Benign providence has supplied the mouse.

Occasionally knockout experiments are easy, and modifying a piece of DNA sequence produces mice that display a grossly altered phenotype. More frequently, however, evolution is not friend but foe. Natural selection has buffered the mice against catastrophe, and although the gene normally plays a role in an important physiological process, a supporting cast is ready to spring into action once it is incapacitated. Modifying the gene produces only subtle differences, whose detection requires knowledge of the biochemical function of the gene's protein and of the details of mouse development. To probe the more recondite relations between genes and phenotypes it is important to understand the ways in which proteins interact in major metabolic processes common to all eukaryotes and the patterns of development that occur in a wide variety of multicellular organisms. We need other organisms in turn to serve as models for the mouse. We need yeast, the fruit fly, and the worm.

So we return to the place from which we started—and perhaps know it for the first time. Promising that the twenty-first century will enlighten us about how yeast works, how flies and worms are made, seemed a disappointment, hardly consonant with the fanfare that accompanies the molecular biological revolution. Our proper study is surely our own genome, in all its three billion glorious bases. Yet if the grail is to work any magic, it can only be with the aid of mundane helpers. Without the model organisms—yeast, nematodes, fruit flies,

E. coli, mice, *Dictyostelium*, and others—the *human* sequence data would be scientifically barren, the grail an empty cup. Model organisms are not an afterthought to current research in human genetics. They are at its center.

Through intricate knowledge of simpler organisms, we shall obtain less detailed understanding of mice and piecemeal conclusions about people. The exact scope of our advances in self-knowledge is unpredictable. We should beware enthusiasts who promise a minute analysis of the human organism—and, equally, naysayers who proclaim that there is no road from sequence data to significant biological results. It would be highly surprising if sequencing genomes failed to provide an enormous amount of important information, both about other species and about *Homo sapiens*. Because we share a vast number of traits with other organisms, studying them can yield important insights about human physiology, human development, and human disease.

What now becomes of the debated goal of sequencing the human genome? Genomic sequencing recommended itself as the route to finding all the human genes. So indeed it may prove. If we have a rich understanding of the smaller genomes of the model organisms, however, the information they provide might be used to guide us into the three billion bases. Even without complete sequencing, we may be able to round up all the human genes whose functions we would be able to identify. Sequencing, even genomic sequencing, is crucial to the biology of the twenty-first century, but we may not need the "grail," the full sequence of the human genome. This is just as well. For, as we shall discover, there are serious difficulties in sequencing anything so massive, and it is encouraging to know in advance that we are able to garner rich returns—even rich returns in human biology—in indirect, pragmatic, ways.

Disappointment may linger. Approaching ourselves from the worm's-eye view may teach us much about the physiological and developmental aspects of ourselves that we share with other living things, but it will not reveal the essence of our humanity. So lofty an aspiration should always have seemed problematic. A point-by-point comparison between the genomes of human beings and chimpanzees (our closest evolutionary relatives) would be interesting, showing us the (probably small) number of places at which new alleles are present and where genes have been rearranged, possibly affecting the control of gene expression. The comparison would resolve some evolutionary questions about the birth of our own species. Almost certainly it would not display the wheel-inventing locus, the language-using locus, the fresco-painting locus, or the genome-probing locus.

People who want molecular biology to answer the question "What is so distinctive about being human?" typically do not have recherché evolutionary questions in mind. They want to know what has made it possible for us to communicate with one another, to compose string quartets, write poems, build cathedrals, harness subatomic sources of energy, and study the interactions of molecules in living things. Nothing in our sequence of A's, C's, G's, and T's is remotely likely to provide an answer. We probably hold that sequence in common, to all intents and purposes, with many other creatures who inhabited our planet during the past few tens of thousands of years. Our human ancestors, even ancestors who roamed the savannahs or huddled in caves, shared with us the capacity for culture. It is conceivable that within the next century molecular biology, developmental biology, and neuroscience will join with psychology and linguistics to show how the capacity for communication, attention, and sophisticated planning relates to features of our brains, how those brains develop, and how the (probably small) genetic differences between people and chimpanzees bear on the typical neurological development of members of the two species. With luck, we shall learn why chimpanzees develop one sort of neural organization and human beings usually acquire another, and how these differences affect behavioral repertoires.

But that will still not answer the ambitious question. We do not just want to explain the *capacity* for culture: We want to know how we have been able to build complex civilizations, while animals whose DNA is very like ours have not. Even if molecular biology, developmental biology, and neuroscience fulfill all their promise, help from other fields will be needed if we are to appreciate how the capacity for culture has been expressed in ramified ways, how (probably small) initial differences have been magnified during the past few million years. Obtaining that understanding will demand the resources of a mass of natural and social sciences, many as yet in their infancy. Exposing "the essence of humanity" is not a task for the biology of the twenty-first century. It is not a task for biology *alone*.

So far we have pondered the value of the sequence-bearing angel's gift. Unfortunately, there is no such angel, and human toil will have to deliver the immense string of A's, C's, G's, and T's. Can it be done?

At present rates of sequencing, generating the full three billion bases is out of the question. Directors of genome centers presently project, with striking unanimity, that within a year they will achieve a megabase of finished sequence per year. The technicians who do the daily work are more cautious, suggesting that the rate will be around seven hundred kilobases. Assuming the more optimistic figure, the labor required to sequence the human genome would be three thousand center years. Perhaps thirty centers with the necessary combination of equipment and expertise could be organized worldwide. So it would take a century to finish the job.

Of course, molecular biologists anticipate that rates of sequencing will increase. They must improve at least tenfold, and ideally one hundredfold, if the project is to become manageable. Sequencing the human genome alone would create an enormous white elephant. In the best of all worlds, we would obtain full sequences for the human genome, the mouse genome, and the genomes of several model organisms: *E. coli* has about 4 Mb, yeast about 15 Mb, *C. elegans* around 100 Mb, *Drosophila* about 150 Mb, the mouse, like other mammals, about 3,000 Mb. The smaller genomes are unproblematic. With envisageable advances in sequencing technology, yeast and the nematode should be finished by the end of the century, the fruit fly within a decade from now. But unless there are dramatic breakthroughs, the full human sequence and the full mouse sequence are likely to remain beyond our reach.

The recent history of molecular biology shows how tasks once deemed impossible can be carried out, at first with difficulty, then by research students, and finally by machines. Optimists about the future of sequencing divide into two faiths. Some put their trust in "Star Wars technology," methods of reading the bases that are quite different from the gel-based methods now in use. Scanning tunnel electron microscopy uses a microscopic needle to "feel" the tiny differences in forces associated with different bases—an ingenious idea that founders at present on the problem of keeping DNA molecules still. Another method aims to snip individual bases from a segment of DNA and let them fall through a laser designed to detect differences in fluorescence, but it has not so far lived up to its early promise. Reluctantly, researchers who hope for sequencing improvements often return to gels and to the majority view that the advances will come by streamlining conventional techniques.

There are several suggestions. Maybe we could use a highly directed sequencing strategy, based on a highly detailed physical map. Unfortunately, converting sequencing labor into mapping labor demands a large amount of preparatory work, including overcoming many difficulties that currently attend mass sequencing. Or we might look to robots to speed up the conventional process. However, so long as there remain recalcitrant bottlenecks (as in the preparation of gels and the analysis of data), the power of automation to boost rates will be limited. Finally, we could always declare that our standards for finished sequence are too high and take a more relaxed attitude about error rates. By doing so we might eliminate much of the redundancy that attends mass sequencing, but welcome relief cannot be purchased at the cost of subverting the enterprise of picking out the genes. If errors are tolerated, can we be confident that the methods of gene hunting will continue to perform reliably?

The challenge is neither so difficult as to be hopeless, nor a matter of conducting business as usual. Perhaps the official rhetoric of the Human Genome Project may exert just the pressure required to force a breakthrough: Like the threat of execution, focusing biological attention on a massive sequencing project may concentrate the mind wonderfully. Fans of genomic sequencing trust that dedicated work today will make large-scale sequencing a trivial affair for our successors, something that the thinkers of tomorrow's theoretical biology can leave to technicians and machines. That is probably too much to hope for. But with a little luck we may piece together an almost-complete, error-tolerant sequence for the human genome, a battered facsimile of the grail.

No doubt it is good to set our sights high. Yet the moral of the tale I have told is that settling for less—when "less" is a set of full sequences of organisms with smaller genomes and a sampling of the human and mouse genomes directed by our understanding of the model organisms—is not settling for *much* less. A decade hence, questions about large-scale sequencing may belong to the remote past, and huge chunks of sequence may be commissioned as easily as biologists now order strains of flies. Alternatively, for all the hoped-for refinements, mass sequencing may still be an intolerably lengthy process, leaving us no practical option except to rely on model organisms for insights into how to sample the human genome. However it turns out, we can be confident that future generations will understand the intimate workings of eukaryotic cells, that they will know vastly more than we do about the physiology and development of worms and flies—and that much of the new knowledge will be applicable to our own species.

In 1900, in the wake of the rediscovery of ideas that Gregor Mendel had formulated over three decades earlier, geneticists anxious to tackle problems of human heredity had two alternatives. One, inspired by the work of Francis Galton (Charles Darwin's cousin), was to continue the statistical analysis of human phenotypes. The other, far more indirect, was to develop a deep understanding of hereditary transmission in other organisms, to expend decades of effort on analyzing fruit flies, molds, corn, yeast, bacteria, and viruses. That indirect strategy paid off handsomely and is the ultimate source of the recent successes of human genetics. We should beware lest contemporary rhetoric celebrating the direct assault on human DNA blinds us to the lessons of history. Like Arthur's knights journeying in the hope of finding the grail, we are almost bound to discover wondrous things—even if the grail itself remains elusive. In ways we cannot yet foresee, fathoming the physiology and development of yeast and worms, flies and mice, will point us to promoting human health.

5

A Patchwork of Therapies

Human bodies are products of millions of years of evolution, which has opportunistically fashioned a collage of the materials at hand, fitting us more or less well for the world in which we now live. Our cells need to copy DNA, to break down some molecules, and to redistribute others, and we share with vast numbers of other organisms the basic mechanisms for performing these tasks that were fashioned by our remote, shared ancestors. Human brains need to send signals to our muscles, our stomachs need to transform various kinds of food, our kidneys need to process waste—but so do the brains, stomachs, and kidneys of other animals, and many of our ways of responding to these requirements are trivial variants of common, even near-universal themes. Breakdowns in the complicated machinery, the sources of disease and disability, can be investigated by looking at the analogues in our evolutionary relatives, sometimes even in very distant relatives like yeast.

Of course what we really want to do is to fix the damage or, even better, prevent the bungled design from affecting performance. Simply singling out the molecules that star in a particular piece of physiological drama, and seeing how the action unfolds, is not our final goal. We want to intervene, to find ways of accommodating the troubles produced by absent, or defective, proteins. Knowledge of how bodies work—or, perhaps, of how minute parts of bodies work—holds out the promise of fixing them when they go wrong.

Forty years after the discovery of the molecular basis of sickle-cell anemia, we know that considerable understanding of mechanisms that misfire to produce a disease does not necessarily increase our capacity to treat it, to prevent it, or even to alleviate the symptoms. Sickle-cell disease occurs in people (most frequently native Africans and those of African descent) who have two copies of an allele that codes for a slightly abnormal version of one of the major constituents of hemoglobin. Hemoglobin is carried around our bodies in red blood cells, doing the important job of delivering oxygen to muscles and organs. To all intents and purposes, red blood cells are simply sacs for hemoglobin molecules; they have no genetic material of their own. People with two copies of the mutant allele have red blood cells containing only abnormal hemoglobin. Under conditions of low oxygen—perhaps when the people are working hard or are at high altitudes—the abnormal hemoglobin causes the red blood cells to become malformed, so that they take on a characteristic "sickle" shape. Rigid and unyielding, the distorted cells become stuck in capillaries, blocking blood flow and preventing the delivery of oxygen. Our understanding of the mechanism has proved very instructive in revealing what will *not* work—and may have saved researchers from chasing attractive wild geese—but it has not so far pointed us toward what will.

The difficulty of dealing with the disease stems in large measure from the peculiarity of the red blood cells. Because they lack their own assembly of genes and distinctive proteins, it is very hard to send normal hemoglobin to them in order to combat the sickling tendency: So perhaps we should think of sickle-cell disease not as a gloomy warning of our future helplessness, but as an anomaly, unrepresentative of the majority of genetic diseases.

Molecular medicine is in its infancy, a bundle of promises with a few encouraging performances. In the last few years, those suffering from various forms of hemophilia have benefited from the production of blood-clotting factors using the techniques of recombinant DNA. Human growth hormone has been manufactured and used to treat some forms of dwarfism and, similarly, people suffering from acute thyroid deficiency (cretinism) have received sufficient doses of the missing enzyme to enable them to develop normally and lead fulfilling lives. As we shall see below, advances in molecular biology have also begun to transform the treatment of cystic fibrosis and of one form of severe combined immune deficiency (SCID). So far these are scattered results, appropriate counters to the pessimism induced by reflection on the apparent impasse with sickle-cell disease, but only tiny indications of a much vaster array of techniques to come. Our first task is to recognize the multileveled possibilities that the new molecular biology may bring.

From the molecular point of view, diseases occur when the body's machinery malfunctions. Any number of things may go wrong. Microorganisms sometimes subvert our inner workings, causing infectious diseases. Other ills can be brought about by molecular intruders from the environment, pollutants in the air we breathe or the water we drink. Yet other diseases seem traceable to our inheritance of molecular mistakes.

For the present, I shall work with a very simple conception of one kind of genetic disease: People who bear a particular assemblage of alleles and grow up in standard environments invariably manifest a disease. "Standard environments" are the circumstances in which people normally live, constituted by the foods they eat, their patterns of daily activity, and a host of other factors, so that this notion of genetic disease must always be thought of as relying on the way of life of a particular group. Canavan's disease, Huntington's disease, cystic fibrosis (CF), and PKU are all genetic diseases in this simple sense, but there are important differences among the examples. Whereas in the first two cases we know of no measures to modify the patient's environment to bring relief from the disease, the adverse effects of PKU can be prevented by altering the diet, and the severity of cystic fibrosis can be mitigated by a growing number of treatments. Genetic diseases challenge doctors to find methods of changing standard environments, to find new ways in which people can live with their rebellious genes.

A naive thought is that genetic diseases have to be treated by modifying some of the genes. Gene modification—more exactly, *gene replacement therapy*—is a tactic that contemporary molecular genetics has begun to develop, but as our examples already reveal, it is only one among many possibilities. Our bodily machinery is sufficiently intricate that we can sometimes find ways to compensate for its aberrant behavior with adjustments elsewhere. Indeed, to announce the main moral of this chapter, if future molecular knowledge bequeaths to us the kinds of interventions we hope for, they are likely to be a motley of techniques, only a very few of which will involve direct tinkering with the genes.

The route from genotype to phenotype is often long and circuitous, and doctors may try to intervene at any of a number of stages. They can try to change the immediate milieu in which the genes act, to deal with the consequences of their activity within cells, tissues, or organs, or to modify the environment external to the patient. Present responses to PKU embody the last approach. We know that the effect of the defective alleles is to interfere with the processing of phenylalanine, so that in standard environments—those in which children receive a normal diet—phenylalanine builds up in the body and does not undergo the usual sequence of reactions that produce tyrosine. Providing a low-phenylalanine, high-tyrosine diet prevents the accumulation and restores normal levels of tyrosine. By changing the environment, we forestall a genetic disease.

Environmental interventions may be imperfect. The special diet for PKU is expensive, socially disruptive, and unpleasant; the costs are small in comparison with the gravity of the condition avoided (mental retardation, usually severe)—although keeping this firmly in mind may require great discipline on the part of patients and their families. Similarly, the standard method of treating beta thalassemia—like sickle-cell anemia, a disease caused by malfunctioning hemoglobin—is to provide patients with frequent blood transfusions, causing them to accumulate large amounts of iron. The excess iron often interferes with normal growth and development, as well as producing damage to the heart and liver. Changing the patient's environment to cope with the effects of defective or absent gene products may introduce new problems that have to be addressed.

Perhaps we would minimize unwanted side effects by coming as close as we can to remedying the fundamental causes? Not necessarily. Interventions at earlier stages of the causal pathway from genotype to phenotype might involve invasive or inaccurate procedures that disrupt other aspects of the body's functioning. Pragmatism and a respect for the nuances of different cases constitute the best therapeutic philosophy. We should appreciate the many possibilities that molecular medicine might offer and be prepared to tolerate solutions that are baroque and indirect. Natural selection does not care about aesthetic elegance in co-opting old parts to new functions. Neither should we.

Some advances can already be foreseen. Even without adding *new* strategies of intervention, increased knowledge of the molecular processes involved in diseases may help us make better use of techniques we already have. Among the least celebrated contributions of contemporary molecular genetics is the ability to engage in "disambiguating diagnosis." Doctors have known for decades that not all cases of the major diseases are alike: One patient will respond to a particular course of medicine, another is untouched, or even harmed, by it. Learning about differences at the molecular level can lead to improved prescription. High levels of cholesterol in one patient are not a matter of diet and exercise but of failure in liver function; the tumors in two patients are built through the action of different genes, and recognizing the differences, we can see in advance that one form of chemotherapy is appropriate for the one, an alternative form for the other. Examples of successful disambiguating diagnosis are likely to multiply in the next decades, and some of them will make the difference between life and death.

Besides better use of old tricks, we can also anticipate novel ways of combating disease, and examples of past breakthroughs indicate possibilities we may hope to emulate. Defective genes typically fail to yield a protein needed for a particular reaction. After finding, cloning, and sequencing the gene, we may be able to discover the protein and manufacture it in large quantities. The obvious next step is to introduce it into the patient's body, specifically to those systems, organs, or tissues that need it for the important reaction. Although this may sometimes prove successful, considerable ingenuity is typically required: In one form of Gaucher's disease, a missing protein causes the buildup of large fat molecules in some cells, producing damaging effects to a number of organs (including liver and brain). Early attempts to supply the protein were frustrated because it was taken up by the wrong cells; after some tinkering, biomedical researchers produced a form of the molecule that binds preferentially to the cells in which it is needed. Moreover, even if the right stuff can be sent to the right place, a program of injections can cause immune system reactions, requiring further tweaking to persuade the system to greet the new protein as a welcome friend and not an alien interloper.

Clearing up the mischief of a defective gene need not involve delivering exactly the protein that is missing. Anything able to perform the same job will do. Sometimes a disease results because a substance that would normally be broken down accumulates in large quantities and can be treated by finding another way to dispose of it. Wilson's disease is caused by the buildup of copper in the liver and brain, ultimately as the result of a missing enzyme. The drug penicillamine binds to the copper and enables it to be excreted, so that patients become "decoppered" and able to lead normal lives.

Just as tests for PKU and Down syndrome belong to an older practice of medical testing, treatments for Gaucher's disease and Wilson's disease are parts of an older molecular medicine; the crucial proteins were identified before the recent explosion of results in human molecular genetics. (However, recent advances have already suggested new techniques that may transform the treatment of Gaucher's disease.) Biomedical researchers hope that as they learn more about the molecules involved in other processes that go awry, these early successes will be imitated on a grand scale. Some diseases will be revealed as involving the loss of a crucial substance, others as caused by the accumulation of damaging biological debris. The aim will be to draw on an arsenal of techniques for producing the allies and disposing of the enemies—and, of course, these techniques are likely to be available to varying degrees, and with varying success, across a wide range of cases.

One approach to treating CF patients shows how insights into the kinds of molecules that clutter the bodies of the afflicted can inspire methods of attack. Because so many people with CF die as the result of a history of lung infections, doctors quite reasonably concentrate on trying to clear away the mucus secretions that invite germs. Discovering that the viscosity of the secretions depends on the presence of DNA that accumulates outside lung cells (it is apparently deposited when the immune system responds to bacterial infections), doctors have explored the possibility of breaking down the unwanted DNA. The enzyme DNAase has the function of "digesting" extracellular DNA in our bodies, and using virtuoso molecular genetics, researchers have designed an aerosol that will deliver DNAase in quantity to the lungs. The results do not turn CF lungs into normal lungs, but the patients who have been treated do show a significant increase in lung function, which translates into fewer infections, less time spent in the hospital, lowered demand for antibiotics, and shorter periods spent at home recovering.

Not all future molecular interventions need be modeled on the achievements of the past. Other scenarios should not be beyond our imaginings. Perhaps we shall be able to treat a number of diseases by intervening much later in the causal pathway. Some autoimmune disorders are brought about because a mutant allele prevents a certain kind of immune cell from recognizing the difference between the patient's own cells and foreign cells. We would prefer to modify these immune cells so that they know their own friends, but as a last resort, we may be able to inject a substance that seeks them out and incapacitates them. Defenders are better dead than mutinous.

There are many more possibilities. When a defective gene fails to yield a protein that would break down a molecule that is toxic in large quantities, we may be able to find ways of modifying the biochemical pathway leading to the noxious molecule. If another gene produces a protein that fortuitously represses the transcription of a second gene, we may be able to deliver molecules that bind strongly to the subversive protein and allow transcription to go forward. Tolerance for tissues may be increased by making molecular interventions in the immune system, allowing greater chances of success with organ transplants. Molecular tricks may disable invading microorganisms, thus preventing or curing some types of infectious disease—a hope that lies behind an intensively pursued line of research on AIDS that has so far met only very limited success. The applications of molecular genetics may be highly indirect.

We are tempted to think of biomedical research as compartmentalized. In the fight against disease, we anticipate attacking head-on. Yet successful strategies often replace one problem with others, and then apply knowledge developed in coping with quite different diseases to resolve the secondary difficulties. This will be the likely shape of things to come. To the extent that the biology of the twenty-first century proves successful in identifying human genes, the structures and functions of their proteins, and the patterns of protein interactions, it will amass an enormous body of knowledge that will be applicable in simplifying, transforming, and solving problems. The result will almost certainly be a hodgepodge of jury-rigged techniques that are used to refine and supplement one another in ways that are initially unpredictable but which ultimately stem from the systematic search for as many human genes as possible and from as much understanding as we can achieve of their effects on the phenotype.

Yet sometimes our efforts at higher-level intervention may be fruitless, and we may be driven to try the most obvious direct assault on genetic diseases. In these cases, mutant alleles begin the sequence of events that culminates in pain or disability—and ultimately we may try to solve some problems at their root by replacing the offending DNA.

Gene replacement might seem to be the only possible approach to some kinds of genetic disease. If an allele that is expressed early in development mutates, there can be widespread consequences of the failure to produce a normal protein. No amount of subsequent mopping up may suffice to repair all the damage. Mutations in early-acting genes probably account for numerous spontaneous abortions (miscarriages), and perhaps also for some of the most terrible hereditary malformations. Yet it is far from obvious what can be done if we determine that a fetus bears one of these genetic conditions, even were we confident that we had identified the causes of the phenotypic problems. All we can do at present is to terminate the pregnancy, and without unprecedented advances in our power to treat fetuses *in utero*, we have little hope of doing better. Moreover, if the mutant genes act sufficiently early, then their irreversible effects may occur long before the genetic condition can be discovered.

All this is true of babies born in the old-fashioned way. In cases of *in vitro* fertilization (IVF) we have access to cells right from the beginning. We can remove cells without causing damage when the fertilized egg has divided just a few times, and we can run genetic tests at this stage; in principle, it would be possible to replace early-acting genes then. There are simpler alternatives. Instead of making heroic efforts to fix major genetic defects, we can select for implantation only those embryos that pass all our genetic tests with flying colors.

How, then, could gene replacement therapy ever be the preferred solution to a medical problem? Apparently, where gene replacement is most needed—in the case of genes expressed early in development—it is either impossible (pretechnological pregnancies) or else redundant (IVF). Appearances are deceptive. The counsel of pragmatism will direct us to include gene replacement therapy in our arsenal of techniques, as something to fall back on when we are baffled in attempts to intervene at higher levels. For example, rather than trying to deliver a normal protein to target cells, we may sometimes find it easier to send some DNA and have the cells manufacture their own protein.

Here too it will sometimes pay to be devious and indirect. Although a patient's condition may result from one or more mutant alleles, the DNA introduced to treat the disease need not carry the corresponding normal sequence(s). Sometimes instead of restoring the ability to make the normal protein, the task will be to inactivate a mutant protein—a job we can do by introducing a gene that codes for a protein that binds to, or breaks up, the unwanted molecules. Or we might want to seek out and destroy some cells, perhaps by introducing genes encoding cell-surface proteins that attract the attention of the immune system. If we could detect cells in the early stages of building a tumor, we might try to deliver genes that would incapacitate them. An earlier moral applies. Large-scale understanding of genomes and their products is likely to yield some straightforward ways of attacking medical problems and perhaps an even greater number of roundabout ways of tricking our bodies into a semblance of normal functioning.

So gene replacement would be a useful weapon—if we could develop it. At the moment, "replacement" is a misnomer: With luck, we can sometimes send some extra DNA to the cells we want to modify. Reaching the right destination is the first major problem. Currently, there are two lines of solution. One mode of delivery proceeds by removing cells from the patient, injecting DNA into those cells, growing the modified cells in large numbers, and finally reinserting them into the patient. The other relies on inserting the desired segment of DNA into a carrier (*vector*), usually a virus that has been specially modified to render it harmless and that is equipped with a capacity to recognize the target cells; the patient is inoculated with the virus and, if all goes well, viruses enter the appropriate cells and deliver the DNA.

Neither method works with anything approaching perfect reliability. Use of the first technique virtually ensures that the desired DNA will reach some cells that need it, but doctors often find themselves caught between two constraints. Removing many cells would risk serious damage to the patient; removing only a small number would probably produce too few modified cells to do any good. The trick is to identify and remove *stem* cells, such as some bone marrow cells, that divide to produce numerous descendants. Unfortunately, only a very small fraction of bone marrow cells (about 0.01 percent) are stem cells, and because of the difficulties of detecting them reliably, attempts to use the technique have had only limited success. Other stem cells are easier to isolate, and in a recent breakthrough one research team has found a way to remove and modify the stem cells that give rise to mouse sperm. Mice today, humans tomorrow. Delivering DNA to human sperm may become possible far sooner than optimists had hoped, far earlier than those with ethical qualms had feared.

The alternative technique, apparently applicable in a broader range of cases, is to use viral vectors. Ideally, we would like to equip the protein coat of the virus with a site that binds preferentially to the cells we hope to reach. Increased knowledge of the molecular biology of various kinds of cells, specifically about their membrane proteins, will probably help us to engineer viruses for particular roles in DNA delivery. At present, however, infecting significant numbers of the right cells is a chancy business.

One important limitation on delivery techniques results from the inaccessibility of the brain—for the trouble in many diseases (such as Huntington's disease, Fragile X syndrome, and Lesch-Nyhan syndrome) stems from what happens in the brain. The most reliable viral vectors—retroviruses—can deliver DNA only to cells that divide, since they rely on the mechanisms for cell division to do their job. Neurons do not divide. Researchers hope to circumvent the difficulty by designing different viral vectors, or by removing and modifying the cells that give rise to neurons, or perhaps by introducing genetically modified cells that can make contact with nonfunctional neurons. As things now stand, however, there are no reliable ways to send genetic material to the brain. While it appears feasible to target some cells of patients with Lesch-Nyhan syndrome (for example, offering relief from some of the gouty symptoms), it is impossible to alleviate the behavioral disorders that constitute the worst aspects of the disease.

For the new DNA to do some good, it must not merely reach the appropriate cells but also find its way into the genome. Elaborate control of this step is impossible: So far as we know, insertion into the host chromosomes is random. This problem is less serious than it might appear. Although we might worry that the new DNA might land in the middle of a functional gene, thus disrupting important aspects of cellular functioning, the sparse distribution of genes in human (or mammalian) genomes comes to our aid. If coding regions occupy about a tenth of our genome, then even if all inserts within coding regions would cause trouble, only a tenth of the cells reached would be disabled. More subtle difficulties arise from the need to regulate expression of the new genes. If we were simply to deliver a segment of DNA without any of the pieces that normally control transcription, then it would very likely land in the long stretches between coding regions, sitting there inert for the life of the cell and doing nothing to ameliorate the patient's condition. The best of all possible scenarios would be to use the mechanisms that normal cells use to control expression of the pertinent gene and insert the new DNA exactly where it belongs, but this is far beyond present technological horizons.

Researchers currently settle for a different solution. They add to the protein-coding part of the DNA insert a control system of its own, something they hope will turn on the gene once it finds a home on one of the cell's chromosomes. But such regulation is extraordinarily difficult. The new control system must not interfere with the cell's own processes, including the regulation of other genes, nor must the regulatory activity be blocked by the internal dynamics of the cell. Just as finding the right cells is currently hit-or-miss, so too is gene expression. All too often, experimental tests on a promising method of gene replacement reveal that very little happens: either the genes do not reach their destinations or they fail to perform once they get there.

One day we may aspire to do what normal cells do—control the exact times of gene expression. For the moment, however, a reliable means of keeping a gene switched on would count as a major success. Ironically, even so crude a capacity is enough to ameliorate some very serious conditions. In some forms of severe combined immune deficiency (SCID), for example, breakdown of the immune system results from the presence of two copies of a mutant allele, so that a protein that is constantly needed is never produced. Children with SCID have begun to be treated with gene replacement therapy; they receive DNA segments that direct formation of the crucial protein and that have a regulatory system which ensures permanent expression. The early results are encouraging: Two girls began receiving gene replacement therapy more than five years ago, and more recently three boys have been given gene replacement therapy since their birth (a period of about two and a half years). In all cases, immune responses have improved, showing an increase in the number of circulating T cells, and so far the only drawback has been episodes of mild, transient rise in temperature after each replacement. Some of the patients have benefited more than others. One of the girls, who was four at the time the interventions began, was given eleven treatments of gene replacement therapy, ending in August 1992. Her immune system now functions essentially normally: Half her circulating T cells have taken up the missing gene, and the cells that incorporate the new DNA have about half the level of functional protein found in her father's T cells (her father, having one normal and one mutant allele, has about half the level of functioning protein of someone with two normal alleles, but fortunately this appears to be a case in which nature is tolerant). The other girl, who began the treatment when she was eight, has a much lower frequency of T cells that incorporate the new DNA (only 1 percent take up the missing gene), and these cells seem to have lives that are shorter than normal. The history of treatment so far underscores both the difficulties and the promise of gene replacement therapy: Inserting new DNA can bring real medical benefits, but present techniques encounter serious difficulties in delivering the DNA to the right places and persuading it to work properly once it arrives.

Another early success of gene replacement therapy has lowered the level of LDL cholesterol in an extreme case of familial hyper-cholesterolemia. A French Canadian woman, currently in her late twenties, suffered a heart attack at sixteen and underwent coronary bypass surgery at twenty-six. Before receiving gene replacement therapy, her level of LDL cholesterol was extremely high, and her level of HDL (high-density lipoprotein cholesterol, the "good" cho-lesterol) cholesterol was low (482 and 43, respectively). After an op-eration in which part of her liver was removed, the liver cells were genetically modified and reinjected into her body. Her cholesterol levels have moved in healthy directions, although they remain highly abnormal (404 and 51, when she is not given special medicine; 356 and 54, when she receives the medicine). These improvements have been stable over eighteen months, and during this period her coro-nary artery disease has not progressed. Gene replacement has not worked miracles, but it has begun to moderate an extreme condition.

In the future, we may hope to find more ways of bombarding the cells of those who are likely to die for the lack of a crucial protein. Moreover, if the problem to be solved is one of disabling cells—as in earlier speculations about the possibility of destroying incipient tu-mors—we can continue to ignore the niceties of timing. Although shooting some DNA into a patient's body, in the hope that it will be taken up by the right cells and will constantly churn out protein, is the molecular equivalent of hitting the body with a mallet, some-times, apparently, a good whack is just what the doctor ordered.

Gene replacement therapy is very young. We may expect it to mature as the systematic exploration of genomes increases our understanding of the internal dynamics of various kinds of cells, as we learn more about the functions of proteins that our genes encode, and as we discover the nuances of gene regulation. Consonant with the therapeutic pragmatism advocated above, we should not fix our hopes on general breakthroughs in solving delivery and timing problems—welcome though these would be—but recognize that the next decades may well bring a number of useful techniques, of various degrees of applicability, that will enable us to transcend current limitations in an unpredictable array of cases.

What *can* be done, either now or in the foreseeable future, is not necessarily what *should* be done. Doctors fiddle with patients' bodies in all kinds of ways, but ought they be allowed to tamper with the genes? A blunt answer turns the question: Nobody objects to efforts at restoring gene products—as when those afflicted with dwarfism are given human growth hormone—so why should we worry about delivering nucleic acid rather than trying to inject a protein? Contemporary discussions usually do not try to mark the distinction between nucleic acids and proteins as a moral line that should not be crossed.

Virtually all of our knowledge of the rival merits of vectors and the vagaries of expression of newly inserted DNA comes from studies on nonhuman animals, most notably the mouse. Quite justifiably, in light of the inexactness of current techniques, researchers have refrained from any large-scale program of injecting patients with new genetic material. In a very cautious and limited way, however, gene replacement therapy has already begun. As noted above, there have been signs of success in treating SCID and hypercholesterolemia, and the recent manufacture of an aerosol to transport appropriately constructed viruses to the lungs promises to bring relief to patients with cystic fibrosis (besides the aerosol that delivers DNAase, there is also a newer spray that aims to introduce the protein that CF patients are missing).

The semiofficial catechism suggests that acceptance of these treatments depends on two main features of the cases: Both involve attempts to cure a disease, not to increase an ability that is already at a normal level; both direct new DNA at somatic cells, not at the *germline* (the gametes, the cells that carry genetic material across the generations). It is worth asking if these are the morally relevant distinctions. As we imagine the future of gene replacement therapy unfolding, should they represent the bounds within which we confine our efforts?

Some possibilities arouse worries about a general license to tamper with the genetic contents of somatic cells. Suppose that the problems of delivery to the brain are solved, so that gene replacement therapy can be used to alter neural chemistry. Perhaps we could prevent the cognitive, emotional, and behavioral problems of Lesch-Nyhan syndrome, liberating boys from the terrible compulsion to bite themselves. Other interventions might prove more troublesome. Our endeavors to tackle a neurodegenerative disease could produce huge changes in mood and emotional response: Someone who is contemplative and morose might become thoughtless and permanently cheerful. Perhaps relatives and friends would wonder if their loved one had survived the genetic therapy. As we have doubts about the effects of electroconvulsive therapy and frontal lobotomy, wondering whether we have released a person from a handicap or replaced one individual with another, so too we might reject some forms of gene replacement because they failed to preserve—and therefore to cure—the person with whom we began.

An even more obvious concern about gene replacement centers on our ignorance of the consequences. Medical interventions often bring uncertainties, and there is no reason to be more fastidious about the risks of injecting DNA than we are in other circumstances where things could go wrong. As with the development of many drugs, claims about the safety of gene replacement will be based on experiments with nonhuman animals. The easiest cases are those in which there is an animal with a similar condition produced by a homologous genotype. Here we have a firmer basis for predicting the effects of gene replacement in people, although differences in background genes or in interactions with the environment might undermine the parallel. When we are trying to treat a disease with important behavioral symptoms, on the other hand, it will be very difficult to forecast the consequences of inserting new DNA into the human body. How would we tell if our prize retrovirus with its special package of genes prevents mental retardation in the mouse (or produces a marked change in the mouse's "personality")?

Sometimes the chanciness of our attempts to deliver and regulate DNA may be reflected in so much variation among experimental animals that we have little basis for judging whether or not the treatment will help human patients. If gene replacement never *worsens* the condition of the animal, however, risks will sometimes be worth taking because the disease is so ravaging. A promising therapy for Lesch-Nyhan syndrome might cause us to wonder about damaging effects to the brain, consequences undetectable in nonhuman animals, but, because of the plight of the boys if we do *not* intervene, we might be justified in taking the chance. Perhaps the current decision to use gene replacement in attacking SCID rests on a similar appraisal of risks and consequences.

If our techniques for inserting DNA become much more exact, many worries may evaporate. Imagine our descendants having the ability to snip out pieces of chromosomes in any cells they like and to replace the excised fragments with DNA of their own choosing (a capability that they have amply confirmed by experiments on nonhuman animals). They could deliver a normal gene to those cells in which a mutant allele causes trouble, confident that the new DNA will be regulated and will perform just as if the patient had been born with a normal genotype. Perhaps there is a small probability of unanticipated side effects, because of some peculiar interaction between the new genetic material and the native genes, but this risk would be very low compared to many now tolerated in medical practice.

Our present crude methods of delivering and controlling DNA, however, involve such large risks of unforeseen damage that germline interventions *and most somatic interventions* are unthinkable. Hence, the distinction between germline and somatic replacement is not crucial for evaluating risks. Indeed, future techniques might be so exact that we would be justified in replacing genetic material either in the soma or the germline. Moreover, germline interventions have one obvious potential advantage, since if the disease allele is replaced in *all* cells, including eggs (or sperm), descendants would be freed from the disease and from the need of somatic therapy.

Perhaps such exact manipulations of DNA will prove impossible—although the history of molecular biology reminds us how quickly the unimaginable is translated into routine. Even if a general ability to replace precisely chosen DNA segments always eludes us, we may be able to manage something similar in important cases. Some genetic diseases—Fragile X syndrome, myotonic dystrophy, Huntington's disease—are caused by the presence of long repeats. Future molecular geneticists may contrive methods to reduce the length of the repeats, for example, by attaching to the patient's DNA molecules that protect a small number of trinucleotides (enough to ensure a repeat within the normal range) and exposing the intermediate stretch to a splicing enzyme.

Whether or not the future conforms to these speculations, we can draw an important moral. One important consideration for gene replacement therapy turns on issues of risk. The questions of risk should be confronted directly, not subordinated to the slogan "Into soma good, into germline bad."

There are *other* important considerations. Replacing genes in the germline affects future people, people who did not consent to those interventions. Shouldn't we be cautious and restrict ourselves to procedures that have consequences only for one person, ideally someone who can agree to what is done? Future people are affected by current decisions, whether or not prospective parents choose to intervene. Often, doctors and pregnant women do the best they can in light of available knowledge to bring healthy children into being. Of course, it is important to allow our descendants to make their own decisions whenever there is a real chance that they would have preferred us not to preempt their choices, but waiting for the consent of the child would frequently be waiting too long. If we are fortunate, molecular medicine will devise gene replacement therapies for Tay-Sachs disease and Lesch-Nyhan syndrome. Whether the DNA is delivered to soma or germline, it would be folly to oppose the therapy on the grounds that the children have not consented to it: If they are to be helped by *either* form of therapy, actions will have to be taken before they can consent—indeed such actions will make it possible for them to consent on other occasions on which their well-being is at stake. Moreover, modifying the germline might bring real benefits to *their* children if the presence of the disease alleles at early developmental stages has debilitating consequences.

Germline interventions sometimes appear frightening because of the possibility of irreversible effects: We imagine the slow decay of our descendants into a population of zombies. Like many bogeys, this is more absurd than terrifying when examined in a clear light. Are we to suppose that everybody is treated (why?) and that nobody notices until the full damage is done? Even more simply, if replacement techniques remain imprecise, then germline interventions (as well as many somatic interventions) are properly debarred by the considerations of risk; if the techniques become exact, then what has been done can be undone through reinserting the original allele.

Talk of gene therapy raises another specter, that of a future in which people "design babies." Thus is inspired the other moral distinction of the semiofficial catechism: Curing disease is a splendid thing, enhancing capacities is not. Hitting a perfectly healthy body with a mallet is not typically a reliable way of improving its functioning, and at present the imprecision of our techniques makes the idea of enhancing our traits ludicrous. Nonetheless, advance thinking about the limits of permissible gene replacement may be worthwhile, if only because the steps we now take, guided by rudimentary powers of manipulating our genome, may set precedents for more striking interventions in the light of increased knowledge.

In a world offering wonderfully exact gene replacement techniques, should we resist the temptation to enhance our own capacities and, those of our children? The formulation already takes for granted a distinction between curing disease and enhancing our traits and, as we shall discover, the concept of disease is more problematic than it first appears. Waiving qualms about the distinction, some conceivable enhancements would seem eminently justified. If we learn how to enhance the operations of the human immune system, reducing vulnerability to infectious diseases and eliminating autoimmune disease, or discover how to prevent the decay of our hearing that normally comes with old age, it would seem unduly pious to abstain on the grounds that we would violate a prohibition on using gene replacement to improve on nature.

One obvious concern is that opportunities would not be open to all. We imagine the privileged fashioning a superrace, while the masses are left to cope with debilitating or degenerative diseases. Injustice can affect the distribution of gene replacement therapies just as it attends the division of other goods. Yet if a future society assures equal access to treatment to all its citizens, if it attends first to urgent health needs before creating the opportunity to enhance capacities, who could complain? Perhaps we might think that some efforts at enhancement are dedicated to securing a competitive edge: We imagine the parents of the future trying to increase the height or the running speed of their offspring; perhaps we envisage them inserting "genes for intelligence" (assuming, of course, that such genes are to be found). If we feel qualms about these kinds of enhancements, it is probably because we imagine them adding further to the arsenal of social strategies available to the privileged. Modeling the future on present affluent societies, we foresee the upper classes investing in a genetic rat race, a race that might even prove self-defeating, while the medical needs of those less well-off go unmet.

Enhancing some human capacities, such as our ability to resist infectious disease, might benefit us all. But a widespread practice of improving on nature—say, by increasing our cognitive powers—could easily have unhappy consequences. A future in which people are frustrated because they cannot lead lives consonant with their abilities is unattractive; one in which society is planned to contain people with enhanced abilities, the alphas, and diminished capacities, the epsilons, is morally repugnant.

We are disturbed by the prospect of enhancing some of our traits, particularly our psychological capacities, because we foresee broad social ramifications with the possibility for catastrophe. Some (but not all) imaginable therapies for enhancing our traits introduce risks of a different kind from those considered earlier. The dangers are not that individuals will experience harmful side effects. Instead there is a real possibility of social conflict, and we have no experimental models from which the risks can be calculated.

Decisions about gene replacement should not be subordinated to global pronouncements or simplistic slogans. Our judgments must be made case by case, reflecting on the needs of the people involved, the risks of the techniques available, the importance of securing consent when we can, the demands of equity and justice, and the social consequences. Taking steps where we can to refashion ourselves and our descendants is not always morally wrong.

Yet once we begin, where will we stop? The shadow of eugenics falls across our present efforts. Although naive horror stories are not difficult to debunk, more serious concerns lurk behind them. For the time being, the limitations of our techniques make decisions reassuringly simple. But we have to consider whether gene replacement offers opportunities we should eschew, whether we are committing ourselves to an immoral enterprise, the enterprise of choosing people. If, as I have suggested, the likely future is one in which our descendants have a motley of therapeutic techniques, set against the background of much broader theoretical knowledge, the use of testing to identify in advance and to prevent the birth of children whose plight medicine could not touch may seem compelling. Should there be limits to such testing? If so, how are they to be charted? Who is to make the crucial decisions?

These important and disturbing questions lurk behind the realistic promise of great improvements in our powers to treat and to cure. Later in this book, I shall try to address them with the care they deserve. First, however, it is necessary to deal with more limited practical issues, problems and possibilities that will be produced by the new ability to gain detailed information about the genes individual people carry.

6

The New Pariahs?

It is hard to hide our genes completely. However devoted someone may be to the privacy of his genotype, others with enough curiosity and knowledge can draw conclusions from the phenotype he presents and from the traits of his relatives. Genetic information circulates among us, sometimes gained unsystematically, sometimes extracted by employers, doctors, or insurers who want to know our family histories. For a few people this is already problematic. Classical genetics, innocent of molecular sophistication, makes it possible to draw reliable conclusions about the probability that these people carry particular alleles—and to decide, on this basis, to deny them the job or the insurance for which they have applied. Perhaps because these instances are relatively rare, perhaps because most inferences from questionnaires license only rough-and-ready judgments, the flow of genetic information has been tolerated. Only a few painful secrets have been exposed to the light.

As genetic tests are introduced and become routine, the fears felt by the few will become the concern of many. Precise, accurate information about each of us will be generated, recorded, shared with people who are committed to help us—doctors, counselors, nurses—and perhaps with others whose interest is less benign. To refuse the tests would be to forego the benefits they might bring; to take them is to make oneself vulnerable. Employers and insurers may demand to know precisely what sequences applicants bear at crucial loci, using their knowledge to winnow out those who are especially bad risks. So would be created a new breed of pariahs, people whose particular combinations of A's, C's, G's, and T's debar them from jobs and from security, people who also bear the burden of high risks of genetic disease.

Faced with the prospect—the genuine possibility—that some of our children or grandchildren might suffer genetic discrimination, it is natural to insist on the privacy of genetic information. How exactly is privacy to be defended? Computer records are notoriously insecure and, as we have already seen, preventing rough, *possibly inaccurate* judgments about people's genes is virtually impossible. Even if batteries of regulations were in place, poised to penalize those caught hacking their way into medical databases, committed snoopers could easily switch their strategy. Carrots and sticks will do the trick, hefty incentives for people who will "volunteer" genetic information about themselves, impossible premiums and no jobs for those who will not.

Emphasizing privacy can easily become a fetish, and it is helpful to reflect on why we value the privacy of information about ourselves. Certain facts about our lives are shared only with our intimates, perhaps not even with them, and we feel ourselves invaded if they are disclosed, irrespective of the ways in which outsiders use the knowledge, irrespective of the attitudes they take toward us in light of it. These things are simply none of their business. We value keeping them private, quite independently of the consequences of publicity. But not all our concerns for privacy are like this. Other pieces of information have to be controlled only because they could serve interests opposed to ours. Making them public would be quite harmless if they could not be used to our detriment.

Checking the flow of the first kind of information is extremely important and, of course, people go to great lengths when they fear they have incautiously confided in an untrustworthy friend. When concern centers on the uses to which information might be put, however, there are two different ways to solve the problem: We can try to prevent the facts from seeping out, or we can place restrictions on what others may do with their illicit knowledge. Monitoring information flow is difficult; it is hard to know just what other people know. Checking their actions is much easier, and imposing regulations on how knowledge can be used has the further advantage of sapping the motivation to pry.

An overromantic view of genetic information comes all too easily. Our genes, the slogan declares, make us what we are. Surely, then, genetic information is bound up with a person's identity, its privacy valuable for its own sake. As we shall discover later, genetic determinism is a myth: People, like other organisms, result from complex causal processes involving DNA and much else besides; our identities depend as much or more on the passage of molecules from the environment as on the constituents of the fertilized egg. Publishing the enormous sequence of a person's A's, C's, G's, and T's would not present him naked to the world's gaze. If that information needs protecting, the reason lies in the uses that others could make of it, not in its intrinsic significance.

Measures to block access to facts about people's genotypes, however heroic, would probably be vain. Luckily, they are not needed. The central problem is to address the possibility of genetic discrimination, already actualized in the small number of cases in which knowledge of family histories serves as a basis for unjust treatment, now threatened on a far broader scale by the advances in molecular biology. Unless we recognize the ways in which genetic discrimination can arise and find effective ways to combat it, the promise of human molecular genetics will be largely unfulfilled. For it is abundantly clear that genetic testing will have little value so long as people fear, quite justifiably, that any information obtained will be used against them.

• • •

During the 1970s, the United States Department of Defense instituted and carried out a policy of excluding individuals with sickle-cell trait—people with one copy of the normal allele and one copy of the sickling allele (people heterozygous at this locus)—from the Air Force Academy, thereby blocking them from the most direct route to positions as commissioned officers in the air force. The policy was officially justified by suggesting that those with sickle-cell trait are at risk for collapsing at high altitudes; in fact, as medical geneticists well knew, the disease occurs only in homozygotes (that is, in those who carry two copies of the sickling allele). Not only were there no controlled statistical studies comparing the performances of people with sickle-cell trait and individuals homozygous for the normal allele, but the behavior of a salient population of well-watched men belied the official claim. American television audiences, including no doubt large numbers of military personnel, observed week after week, season after season, as a group of athletes, roughly 7 percent of whom were known to bear sickle-cell trait, assailed one another, sometimes at Mile High Stadium in Denver, without revealing any difference in propensities for collapse. The high incidence of sickle-cell trait in the National Football League and the formulation of an exclusionary policy on the basis of flimsy evidence very likely have a common explanation: In the United States the sickling allele, and thus both sickle-cell trait and sickle-cell anemia, is most prevalent among Americans of African descent.

People who make use of technical information are bound to make mistakes. Sometimes even the most authoritative scientific findings are incorrect, and though application of the findings may cause harm, those who rely on them cannot be blamed. Doing the best they can, they consult the most prominent experts, taking pains to confirm that the results are well established and that they allow the proposed applications. Apparently, the Department of Defense was not quite so scrupulous: The genetics and physiology of sickle-cell disease and sickle-cell trait had been extensively studied for thirty years before a lawsuit forced the cancellation of the ban. Not all ignorance is culpable—but not all is blameless.

Any legitimate use of genetic information should guard against mistakes that will misclassify people, thus causing them wrongful harm. Infallibility is too much to ask for; responsible application of scientific knowledge is not. Widespread misunderstandings of genetics can breed all kinds of errors: Perhaps sickle-cell trait is confused with sickle-cell disease, perhaps the fact that two doses of the sickling allele is bad is taken to entail that one dose must be half as bad. More subtle mistakes can easily occur. Lumping people together, without regard for environmental differences that are known to be relevant, may generate misleading statistics that portray the risks for some individuals as being much higher than they actually are.

Whatever the merits of using genetic information in particular contexts, in calculating insurance premiums or in evaluating candidates for a position, all such uses are subject to an informational precondition. If genetic information is to be used to classify people, and to treat the classes differently, then those applying the information are responsible for understanding how the variation in genotypes bears on phenotypic traits. Specifically, they must be aware of the different combinations of alleles that can occur at the locus (or loci) of interest, the range of phenotypic variation for each combination of alleles, and for each such combination, the probabilities that people with that combination will display each of the phenotypes, differentiated insofar as is possible according to environmental variables known to be relevant. When powerful institutions use genetic information but violate the informational precondition—as the Department of Defense apparently did—injustice has been done, and those harmed should be compensated.

Suppose, however, that genetic information was thoroughly digested, applied with scrupulous care, and indeed, that the resultant classifications were entirely accurate. Some people judged high risks for future disease or disability might be denied insurance or jobs in consequence. The judgment would be correct: These people are indeed at high risk. If they find themselves uninsurable and unemployable, forced to the margins of their society, that is unfortunate—but is it unjust? When, if ever, does differential treatment based on accurate genetic classification become genetic discrimination?

To find answers to these important questions, it is essential to distinguish various contexts in which genetic information can be applied: in fixing insurance premiums, in evaluating job applicants or employees, in everyday social relations. I shall examine these contexts sequentially, starting with a problem peculiar to the United States.

Other societies in which genetic information is likely to flow freely, the affluent democracies of the first world, are already committed to providing health care for their citizens. Americans, meanwhile, continue to debate the need for universal coverage, often touting the virtues of healthy competition among private insurers. In the battle for market shares, how much genetic information should insurers be allowed to use?

Insurers can make a case for fighting with the gloves off—and, in doing so, may win the applause of consumers who do not want to subsidize the premiums of others. After all, they may suggest, underwriting is the business of adjusting premiums to risk. Genetic information allows a more accurate assessment of risks, and consequently a fairer allotment of premiums: Those who are likely to claim more have to pay more; present mistakes, born of the use of inaccurate information, will be rectified. Moreover, if potential consumers can use information denied to suppliers, they may find it to their advantage to act in ways that will ruin the insurance market. People at very high risk can buy extra coverage, those at low risk may not bother to insure themselves at all, and without the ability to distinguish the two, insurers will discover that claims outrun premiums.

These arguments are perfectly general, applying to insurance of any kind, and there are certainly circumstances in which they have considerable force. Imagine two families, each of which owns some valuable jewels. One family is generous, hospitable, unsuspicious, rather careless, and somewhat absentminded. The house is open to a wide variety of people, the door is often unlocked, and although the diamonds are hidden, family members rarely check that the hiding place and its immediate surroundings are as they should be. The second family is less gregarious, is neat, punctilious about locking doors and checking on valuables, fussy about the moral probity of its social circle. Each family wants to insure its jewels. The insurer knows the relevant facts about the habits of the two families and recognizes that the second family does a better job at protecting its valuables. Is there any reason not to use that information? Would it be unjust to ask the first family to pay a higher premium? Would it even be unjust to set their premiums so high that the cost would be ruinous, or simply to refuse insurance coverage altogether?

Matters are different when the topic is health (not family valuables) and genetic information (not facts about family habits). Health plays a special role in human lives. Requiring cripplingly high premiums of people—even, in the limit, denying them access to health insurance, thus greatly affecting their opportunities for treatment, prevention of disease, and maintenance of health—curtails or excises a capacity fundamental to planning their lives and attaining their goals. It is also pertinent to distinguish facets of behavior that can be changed—as the first family could amend its warmhearted but unfastidious ways—and conditions people are born with. Demanding a ruinous price from people who could join the class of those offered a cheaper bargain, if they chose to do so, is one thing; penalizing people for characteristics they cannot change is quite another.

The disanalogies reveal why the insurers' argument breaks down in general and why it does not apply in the realm of medical insurance. Democratic societies are grounded in the rights of citizens to fashion their lives as they choose, so long as they do not infringe on the equal rights of others. When the actions of one group would deprive members of another group of *any* coherent opportunity to shape their lives, there is a duty to protect those who are threatened. Some social thinkers believe that this principle has wide application, that there are numerous ways in which existing social arrangements cramp the possibilities for poor and disadvantaged citizens to form satisfying projects, and in consequence, that serious reforms are needed to change the environments in which many children develop. Their concerns will surface at the end of this book. For the moment, however, I shall focus on what I take to be the least controversial application: In whatever way we choose to live our lives, the ability to maintain our health is crucial. Without it we cannot plan for the future, cannot set life goals that we might reasonably expect to attain. Unless there are special circumstances, a just society cannot allow insurers' aspirations for a more efficient, or more profitable, business—or consumers' zest for cutting their premiums—to take precedence over the provision of so fundamental an ability.

Occasionally the circumstances are special. Conflict between groups may result from decisions that people have made freely in the past, in full awareness of the consequences, so that they are now vulnerable to the loss of something important for the shaping of their lives. Perhaps higher premiums for health insurance are justified when the applicants have voluntarily overindulged all the wrong habits, knowing the medical effects—although the fashionable war on tobacco, alcohol, and fatty foods might itself be charged with Puritanical excess.

Because the boundaries of free choice and foreknowledge are hard to identify precisely, it will not always be clear when someone has forfeited the right to protection. Issues about using genetic information to set insurance premiums do not, however, call for locating an elusive line of demarcation. A woman born with a mutant allele that confers high probability of cancer should not be declared uninsurable. If, no matter what she had done, she would still have been at higher risk, then the duty to protect her against loss of a fundamental resource, the ability to maintain her health, cannot be canceled by holding her responsible. But should she be forced to pay higher premiums than her more fortunate sisters?

Complete adjustment of premium to risk would surely lead, in a significant number of instances, to imposing costs that would prevent the applicant from obtaining health insurance, breaking the commitment to ensure the possibility of coverage for all: American insurance companies have already offered intolerable choices to hardworking parents whose children have been found to carry CF mutations or the long repeat associated with Fragile X syndrome. It would be perverse to insist that no applicant be declared uninsurable but allow insurance companies to fix premiums so high that they achieved the same end. There are more modest proposals. Perhaps genetic information could be employed to demand higher premiums of those born with unlucky alleles, provided that the difference was not so great as to debar people from coverage.

We should resist the urge to penalize the unfortunate and reward the lucky, even in its more muted forms. Medical insurance is best viewed as a scheme in which each of us participates to ensure ourselves a vital resource. Risks are thrust upon us by the genetic lottery, and in a just society those risks would be irrelevant to the costs of coverage. One of the great metaphors for affluent meritocracy is the image of a race for attractive prizes, in which the contestants are to have (approximately) equal opportunities. Societies that tolerate conditions in which some people are declared medically uninsurable effectively insist that those bearing unlucky alleles must run with extra burdens. Nor is it fair to extort a price for removing the original burden that effectively introduces new handicaps. If we are serious about equalizing opportunities, then those who are both poor and genetically unfortunate should be offered medical coverage at *lower* rates than their fellow citizens. The pertinent variable is not the degree of risk, but the ability to pay.

The most obvious social arrangement for implementing these moral principles is to eliminate medical insurance entirely, developing a system in which care is available to all citizens and funded through progressive taxation. Especially in the United States, plans of this sort are often distrusted because of a faith in the power of economic competition to produce better service and a correlative concern that centralized bureaucracies prove wasteful and inefficient. Perhaps the worries are well founded, and affluent democracies would be well advised to try to combine the benefits of private service in a competitive market with the fundamental commitment to ensuring health care for all at affordable rates. However, the principle that all citizens should have health care coverage at costs determined by their ability to pay is indeed fundamental, and we should not be tempted to compromise it because of panegyrics to the free market. Justice requires that we do not secure whatever advantages accrue from competition by sacrificing the well-being of those who have already been unlucky in the distribution of genes and the distribution of wealth.

Although these simple thoughts have often been taken for granted throughout much of the first world, they remain controversial in the United States and are becoming more disputed elsewhere. Enormous efforts have been made to hide the elementary arithmetical fact that if large numbers of the working poor are to receive the health coverage they currently lack, then their more affluent fellow citizens must pay extra to continue the same level of benefits. Feats of rhetorical legerdemain attempt to block the question, Why should I pay for the health care of others?, but the right strategy is surely to face it directly. Genetic risks are distributed unequally at conception through the quirks of meiosis and fertilization. Economic assets are also dealt unequally, largely as a result of the accidents of birth. Once economic status and genetic endowment are known, it is easy for the winners in the pertinent lotteries to chafe at the idea that they owe anything to the losers. But we live at a time that permits an interesting, and realistic, thought experiment.

Imagine that we are on the eve of the great unveiling. Tomorrow the genetic truth about each of us will be known; all the important disease loci have been sequenced. Roughly 10 percent of the population will discover that they carry allelic combinations placing them at significantly higher risk for some disease. The citizens divide into two groups, which will manage health care coverage in different ways. One class consists of those who will assign medical premiums on the basis of risk, the other comprises people who will set medical premiums independently of risk (and dependent on economic assets). Tomorrow, when the genetic facts are known, 90 percent of those in the first group will be somewhat better off than the members of the second group; the remaining 10 percent will be significantly worse off, and some of them will even be forced to live out the rest of their days without health coverage, since they cannot afford the premiums required of those at high risk. Today, you must throw in your lot with one of the schemes, but you are free to choose either. Which group will you join?

Knowing your assets, aware that it would take extremely bad luck for you to be unable to pay the premiums demanded of the genetically unfortunate, perhaps you might decide to gamble, risking a small chance of being bereft of health care for the satisfaction of lower costs (savings obtained by not having to bear the burden of others). Any such temptation can be reduced by changing the conditions of the thought experiment. Suppose now that there are not two groups, either of which you can join, but two proposals for the way in which the nation's health care will be run. Depending on the votes cast by you and your fellow citizens, one of the two systems will be instituted and will govern health care coverage for the descendants of the current population. As you contemplate the chances for your children and grandchildren there are now two foci of uncertainty: You do not know how they will fare in the genetic lottery, and you cannot be confident of their economic circumstances. How should you vote on their behalf? By trying to free them from the (relatively small) tax of paying to support the genetically unlucky, risking possible disaster if they do badly in both the genetic and economic lotteries? Or by ensuring that however things turn out, they are assured health care they can afford?

Unlike many similar thought experiments, the scenarios just envisaged represent parts of a contemporary predicament. The arrival of routine genetic testing really will amount to a great unveiling, and if insurers are allowed to use genetic information in adjusting premiums, the burden of paying large (if not impossible) sums to obtain health care coverage will afflict numerous Americans who never thought of this as their problem. As the United States has debated health care, maintaining the freedom to choose doctors and preserving access to high technology have seemed primary because most citizens have regarded the possibility of being deprived of coverage as a concern only for other people—the indigent, the undeserving, the feckless. Yet this is no longer true in a world where genetic tests are widespread, and genetic information is available to insurers. There have long been many excellent reasons for viewing a system of health care coverage as a cooperative venture in which the fortunate help to support those who have been victims of circumstance. Because genetic testing can make victims of us all, its advent may help to undermine the glorification of self-sufficiency that has so far blocked appreciation of those reasons. After the great unveiling, any of us—or, more probably, any of our descendants—may be vulnerable.

It is important to come to terms with a second part of the insurers' case. If people were guaranteed the opportunity to buy as much insurance as they pleased, paying premiums at the same rates, or at rates based on ability to pay, then individuals who discovered that they were at high risk might apply for higher coverage, while those to whom genetic tests brought good news might settle for less. Fears of collapse in the insurance market would be well founded.

This problem can be solved by distinguishing that basic level of health maintenance all people need to pursue their projects, whatever those projects may be, and possible supplemental coverage. Genetic information ought to be irrelevant at the first level, and the costs of care should be spread so that all needs can be met. At the second level, where citizens may choose to extend their coverage, earlier arguments no longer apply, and it is quite reasonable to permit the use of genetic information in fixing premiums. Of course, how exactly these levels are set must depend on the state of medical technology and on the resources of the society. Yet it would be wrong to undercut the arguments for universal coverage by providing only the most skeletal services to all citizens. In affluent societies with the power to employ a wide variety of tests, preventive techniques, and treatments that can allow citizens to conceive and pursue projects over a significant life span, the basic imperative to make available those resources that are crucial for self-development will oppose any such stripping of universal coverage. Perhaps supplemental insurance will cover a few experimental procedures or interventions too expensive to offer on a grand scale, perhaps heroic measures for slightly extending the lives of the elderly. Just as there are differences between luxury hotels and plainer lodgings, some insurance plans may offer comforts that go beyond the purely medical needs. But a two-tiered approach should ensure that all citizens have all they need for shaping their lives, while guarding against the threatened ruin of the market.

Although the question of using genetic information in setting health insurance premiums is of particular concern for the United States, all the affluent democracies will confront other insurance issues. Facts about applicants' genotypes would be useful for providers of life and disability insurance. We might assimilate these contexts either to the discussion of health insurance or to the case of the family jewels. Which, if either, is the appropriate model?

Both disability and premature death are threats to the completion of life plans and projects. People who insure against disability seek to prevent a future in which, for all their prudent saving, resources are insufficient to enable them to complete plans they have carefully made, projects that have been central to their lives: Disability thus seems to belong with health and to fall under the scope of the arguments just presented. In a less obvious way, similar considerations might apply to life insurance, for an important part of many people's lives consists in advancing the well-being of loved ones. Even though you may die prematurely, the success of your life is contingent on what happens afterward, on whether those you have cherished continue to thrive. Moreover, an inability to provide against disaster would tinge the present with insecurity.

Concrete examples are commonplace. Imagine a couple with two young children. One spouse, for whatever reason, has no opportunity for working for more than modest pay. Both have been prudent and industrious, providing opportunities for the children, buying a house in a district in which they have confidence in the schools, and the family is now happily established in its social and educational setting. If the primary wage earner were to die prematurely without an insurance payment to the survivors, the family would be forced to make dramatic changes. It would be uprooted from its surroundings and exposed to all the pressures from which, with much forethought and hard work, it has so far been protected. Unfortunately, the wage earner bears a combination of alleles conferring high risk of early death. Allowing insurers to use genetic information in setting premiums for life insurance would make it impossible for the family to purchase coverage at rates it could afford, preventing the couple from staving off the foreseen calamity.

Poignant though such cases undoubtedly are, the case for a system of guaranteed life insurance, parallel to the two-tiered approach to health coverage advocated earlier, is less clear-cut. Because the purpose of life insurance is, almost always, to enable people to secure for their loved ones a range of opportunities which at least some members of society, often a significant number, do not enjoy, considerations of justice pull in different directions. People who have struggled to build opportunities and security for their children deserve sympathy and support, and genetic conditions that are matters of accident, not merit, should not undermine their ability to protect what they have achieved. Yet what they aim to preserve is a level of well-being greater than that enjoyed by members of other groups, many of whom are not responsible for their lack of opportunities. Comparing the vulnerable family with other relatively wealthy households, it appears wrong to make a distinction on the basis of genetic luck, permitting some to guard against the destruction of all they have worked for and depriving others of that option. On the other hand, most affluent democracies contain people, equally concerned about the futures of those they love, who lack the chance to create the kinds of opportunities that are widely desired: Their children are stuck in dangerous, depressing, and unhealthy environments, without access to stimulating schools. A system of guaranteed life insurance would prevent industrious, relatively well-off families from being plunged into similarly bleak circumstances simply because the wage earner happened to do badly in the genetic lottery. It would ignore the fact that other families—significant numbers of families in the United States—experience discomfort and cramped opportunities because of the accidents of birth.

The earlier analysis of the use of genetic information in health insurance rested on a principle: If something is needed to enable people to carry out any life plan, they should be protected against actions that would deprive them of the necessary resource. Unlike access to health care, the ability to secure the well-being of descendants is not required for executing *any* project that people may set for themselves, but it is surely very important to most of us. Can we apply the principle to conclude that there should be a scheme guaranteeing the possibility of purchasing life insurance, at reasonable rates, independent of facts about the applicant's genotype? Not quite. The envisaged scheme would not provide equally for all citizens. It would certainly be a step in the direction of greater justice—some people would not be impeded by genetic conditions, for which they are in no way responsible, from obtaining something of central importance to their lives. If we honor the underlying principle, however, we might feel compelled to strive for a more far-reaching policy, one that would be equally sympathetic to those whose prospects are clouded by the unfortunate circumstances of their birth.

The issues raised here prefigure an important debate that will occupy us at the very end of this book. To what extent should just societies provide all citizens with the resources that all need to fashion and carry out significant life projects? If we believe that the provision of medical care is one necessary part of remedying the accidents of birth, are there others? Some people, aptly thought of as idealists, will suppose that the medical issues are just one prominent manifestation of deep social inequalities, that regulating insurance is a patched and partial solution, and that justice requires us to attend to the ways in which both genes and social circumstances deal good luck or misfortune. Others, pragmatists, will recommend that we do the good we can by protecting people who "play by the rules" but are victimized by their genes and by not allowing their welfare to be hostage to sweeping policies aimed at problems we have little chance of solving. Both can agree that something should be done to control the uses of genetic information by the providers of health, disability, and life insurance. They differ in their views about what else we are required to do.

Carrying alleles that place one at high risk for disease is already a burden. The burden becomes heavier if insurers are allowed to demand higher premiums from people whose genetic conditions put them at risk. It is increased still further if employers are allowed to use genetic information to deny work to the genetically unlucky. If genetic differences are ignored in the contexts of health and disability insurance, one important motivation for using genetic tests in the workplace can be set aside. Employers need no longer fear ruinous expenses if they take on people with genetic predispositions for future disease, for the costs of insuring those people will be no greater than those of insuring other workers. Failure to achieve justice in matters of health and disability insurance begins the process of creating the new pariahs. Conversely, if the steps recommended earlier have been taken, the threat of being sentenced to permanent unemployment is greatly reduced.

However, the danger is not completely dissipated. Although genetic tests are currently moderately expensive (the costs are usually of the order of a few hundred dollars), employers might find it worthwhile to administer them to identify people whose productivity would likely be curtailed, to forestall costly lawsuits from employees affected by substances released in the workplace, or to avoid being forced to clean up the workplace environment. When jobs are sufficiently scarce, potential employees might even have to bear the costs of testing. Production of a clean bill of genetic health might become a condition of application.

Although many people resent their jobs, the fear of being declared permanently unemployable runs very deep. Work may be for those who don't know how to surf (or ski, or fish, or whatever), but the money for surfing (skiing, fishing) would be unavailable without it. Tomorrow's adolescents may discover that contingencies of nucleotide sequence doom some of them to lives of idleness and poverty. A simple proposal for avoiding this outcome would be to ban the use of genetic information in employment situations (and it seems that the United States Equal Opportunities Commission has found this simple suggestion sufficiently attractive to bar employers from demanding genetic tests that could disclose potential health problems in applicants or employees who are currently healthy). The proposal might be justified by the same kind of reasoning that has figured previously: The ability to work is central to developing a satisfying life—because work brings the means (cash) for carrying out plans and projects—and people should not lose that ability because of accidents beyond their control. As in the discussion of life insurance, applying the principle would threaten to plunge us into a much broader social debate. But in any case, the principle does not apply so straightforwardly when we switch from insurance to employment.

Decisions to assign someone a particular job may have profound effects on the well-being of others. When the United States Department of Defense drew up its misbegotten exclusionary policy, a readily comprehensible rationale (irrelevant in light of the biological facts) was probably lurking in the background. Military officials were concerned, quite properly, that pilots likely to collapse at high altitudes would endanger not just themselves and costly pieces of equipment, but other crew members into the bargain. In fact, even if there had been a small but significant probability of collapse, the worry would still have had force. When there is *accurate* information to the effect that bearers of certain combinations of alleles have a nonnegligible probability of behaving in a specific way in a particular environment; when performance of the job requires being in that environment; when behaving in that way in that environment would endanger the health, or the lives, of others; then genetic information is relevant to deciding who should be given the job. Banning the use of genetic information altogether is too simple.

Decisions about how to use genetic information, even when it is pertinent, can be extremely delicate. At least one person working as an air traffic controller in the United States carries the long repeat for Huntington's disease. Should the flying public be protected by forcing him to retire before the symptoms of nervous degeneration begin? Or should we rely on him and on his colleagues to pick up the early manifestations of the disease—as we trust elderly drivers to decide when they should move over to the passenger side? In cases like this, the conclusions to be drawn from the genetic facts are hotly debated, but there are other instances in which the import is very clear. When the risk is that a symptom will manifest itself *suddenly*, with grave consequences for others, the genetic information cannot be set aside in favor of close monitoring of the phenotype.

So there are at least some circumstances in which employers can legitimately use genetic information about applicants or employees. Are there others? A natural thought is that the use of genetic tests is not necessarily discriminatory. Facts about applicants' genotypes might bear on their qualifications for a particular post, and no injustice would be done if employers gathered those facts and made their decisions in light of them. Nor would this procedure inevitably lead to the creation of a class of the permanently unemployable: Being unqualified for a particular job does not mean that one is unqualified for *any* job (although, as we shall see below, when job opportunities are restricted, the point of principle may be meaningless in practice). When symptoms are already present, when the person already has difficulty with the tasks required for any employment, genetic tests are irrelevant and banning them would do nothing to improve the applicant's job prospects. Genetic information is only useful for employers who want to know the chances of future disease and disability. Even if those chances affect qualifications for some positions, they do not do so in general. In many occupations, applicants are rightly judged on what they can do in the short term; employers can cope with the future when it arrives.

Genetic tests might disclose two quite different kinds of facts about applicants: their risks of acquiring various debilitating diseases, whether or not they work in the environment the employer provides (general health risks), and susceptibilities to diseases that might be triggered by substances emitted in the workplace (workplace-specific risks). When are general health risks relevant to qualifications? Assuming that insurance issues have been resolved, so that hiring the genetically unlucky costs no more in premiums, general health risks would affect a candidate's qualifications only if they signaled a likely loss of productivity. Genetic tests might show, for example, that a potential employee is at high risk for a neurological disorder that would inevitably curtail the person's career.

Likely losses in productivity are occasionally relevant to the applicant's qualifications—but not always by any means. The clearest instances are those in which employers have to provide support for a long period of training, during which the employee is relatively unproductive, so that training a replacement would be costly. Here, knowledge about the probable length of the posttraining period might be directly relevant to assessing an applicant's qualifications. Imagine an extreme case in which it is very unlikely that someone with a particular genotype will be able to function in the job for more than a year after an initial expensive training period of five years. Even here it is not the *only* relevant information. When the length of the training period is negligible, however, or when the probable time of employment is affected by other factors—as when there is a rapid turnover of workers quite independently of decline in health—facts about genotypes are irrelevant to evaluating qualifications. Given the uncertainties that usually attend the course of employment, it would be foolish and unjust for employers to give much weight to genetic information, ignoring the other criteria, such as aptitude and personal character, on which they normally rely. Responsible employers worry about future absenteeism and the possibility that experienced workers may leave, but they should evaluate candidates by considering the whole spectrum of traits that bear on these outcomes, not simply focusing on what they can garner from genetic tests. Only in the most extreme circumstances, when it is imperative that the person appointed now be able to see a particular long-term project through, might genetic information about general health risks be a major factor in a just decision.

Testing job applicants for workplace-specific risks may easily seem more benign. Unlike those whose genes dispose them to acquire a debilitating disease, no matter where they work, people with genetic susceptibilities to diseases triggered by particular substances appear to benefit from being prohibited from workplaces where the substances are present. Nevertheless, employers who plan to administer tests for workplace-specific risks are not moved entirely by altruism. Regulatory agencies are often looking over their shoulders, scrutinizing the rate at which workers become sick. Nor are the beneficiaries of the screening always grateful. Despite the fact that an environment is unhealthy for them, they may be driven to seek work in it because they see no alternative. Use of genetic information may not simply debar them from a particular job, albeit a job that would imperil their health, but render them effectively unemployable.

Extreme examples are easy to adjudicate. At one pole are situations in which employers could easily bear the costs of modifying the workplace environment so that genetic differences no longer affect risk: In the new clean plant nobody runs a high risk of acquiring the pertinent disease. At the other pole are cases where potential employees have equally good options in environments that do not increase their risks of disease. Only irresponsible or ignorant employers fail to amend the workplace when it is economically feasible for them to do so; only perverse, or misinformed, applicants choose jobs that raise their risks of disease and disability when equally good alternatives are available.

Troublesome conflicts arise when either no way of modifying the workplace is known, or the only known ways are so expensive that they would bankrupt the employer, and when the applicants are bereft of other job opportunities. There are numerous small towns throughout the industrial (and postindustrial) nations in which there is effectively one employer for the local population, benefiting from a captive workforce, engaged in a manufacturing process that releases substances to which some people are particularly sensitive. From early adulthood on, the locals may be committed to staying: Their entire system of social and family support, which makes their lives go smoothly, would break down if they were to leave. So they apply for work at the only place in town. If the manufacturing environment contains substances toxic to people with their genotype, they are caught in a cruel bind. To keep the recorded disease rates low, the employer would like to use genetic tests to screen them out. The workers have no wish to fall ill or die young, but feeling they have no choice, they prefer to take the gamble.

Barring employers from using genetic information is hardly a humane response to this recurrent predicament—although it may be marginally better than condemning members of a captive workforce to a lifetime of unemployment. The conflict results because there is no feasible way to modify the workplace and no alternatives for the job applicants. Even if nothing can be done about the former, it is possible to loosen the latter constraint. Creating new job opportunities, possibly accompanied by retraining programs, would release the workers from their bind. Especially in places where employers benefit because they are the only game in their respective towns, it would be appropriate to demand that they bear a significant proportion of the costs of job retraining and job replacement programs.

Very often the sharp distinctions that have so far figured in the discussion will be blurred. Opportunities for other jobs will not be entirely lacking: Workers could go elsewhere by accepting a significant cut in pay. Employers could reduce the level of toxins in the workplace to a greater or lesser extent. Resolution of the more complex situations should be guided by the same sensibility that governs simpler instances—concern to free workers from a cruel choice—although the details of evaluating obligations will often be messy.

Analyzing situations in which levels of toxic substances can be reduced by varying amounts, at varying cost, should proceed by considering two questions: How do the risks for people bearing the different genotypes involved depend on the level of substance present? How do the costs of clean-up depend on the extent of the reduction? Different answers will apply in different cases. Consider the possible situations represented in Figure 6.1. If the facts are accurately represented by (a) and (c), it would be possible for the employer to modify the environment in a way that makes genetic testing irrelevant: The differences in risk between people with genotype A and genotype B would be effectively eliminated if the level of the toxic substance were reduced to two-thirds of the present value, and the cost of the reduction would allow the employer to undertake it without financial ruin. On the other hand, if the relative risks are as shown in (b), then, whether or not the schedule of costs accords with (c) or (d), nothing the employer can do will eradicate the difference in risks. Here, bearers of the lucky B genotype are not at high risk for disease and disability, even if the toxic substance is at its present level; on the other hand, those with the A genotype are in great danger unless the toxic subtance is reduced to a level so low that it would be economically impossible (perhaps technologically impossible) to attain. Here, genetic information might appropriately be used, but, in ac-

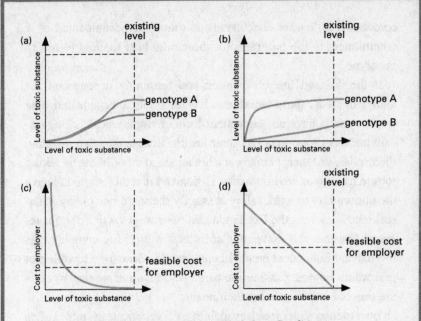

FIGURE 6.1. *Different Workplace Conditions*

Diagrams (a) and (b) show purely hypothetical (but perfectly possible) relations between the effects of different levels of a toxic substance on people with different genotypes. In (a), the existing level of the toxic substance creates a significant difference in the risks for people with the two genotypes, but if the level of the substance were decreased to about two thirds of its present value, that difference would disappear. In (b), by contrast, the differences in risk persist unless the amount of toxic substance present is very small indeed.

Diagrams (c) and (d) show how the cost to the employer varies with the level to which the toxic substance is reduced. In (c), reducing the amount of toxic substance present is initially cheap and becomes progressively more expensive. In (d), the costs of reducing the level by a particular amount remain the same, irrespective of the level currently attained.

Given the hypothetical assumption about feasible cost, it is clear from comparing the graphs that if (a) represents the relations between toxic substance level and risk for the two genotypes, then, given either of the economic scenarios in (c) or (d), it is possible for the employer to reduce the level of toxic substance so that bearers of the two genotypes are at similar low risk. By contrast, if either (c) or (d) represents the economic facts, and if the risks are set as in (b), then whatever the employer does, bearers of the A genotype will be at much higher risk than bearers of the B genotype.

cordance with earlier conclusions, it should be accompanied by a commitment to job programs for those who bear the less fortunate genotype.

If the demand for job creation and retraining depends on acknowledging a captive workforce, it is important to be able to decide when workers have no "equivalent form of employment." A significant number of workers risk their health because they can maintain themselves and their families at a higher standard of living by taking jobs in hazardous environments. To demand that they should receive the opportunity to work safely at exactly the same rate of pay is unrealistic. To accept the likelihood that their wages will be so far reduced that they will have to change their way of life completely is too harsh. "Equivalent employment" should allow for a lowering of real wages without making the bargain so unattractive that workers are coerced into sacrificing their health.

Eventually, widespread availability of genetic tests may soften the hard choices. If a factory is the chief employer in a town, if its manufacturing process releases chemicals to which individuals with certain genotypes are sensitive, if no way is known to lower the frequency of emission, then the best solution may be to ensure that tests are administered while children are still in school so that they can be guided into alternative forms of work. Blanket opposition to the use of genetic information in employment decisions is a natural response to the worry that employers will substitute genetic testing for costly attempts to clean up the workplace. The interests of workers are best served, however, by insisting that environmental modifications be undertaken where it is feasible to do so, that genetic testing be accompanied by the creation of job opportunities, and that genetic tests be given sufficiently early to enable people to avoid choosing between their health and their economic well-being.

Besides the definite harmful effects of losing insurance coverage or being denied a job, there are more nebulous consequences that prompt fear of the release of genetic information. Will people with particular genotypes be stigmatized, bearing a scarlet *A* at some critical point in their nucleotide sequence? At first glance, the worry seems unfounded, for it appears unreasonable to scorn those who carry unlucky alleles, since they are in no way responsible for their genotypes. Members of groups currently stigmatized—American homosexuals, for example—have sometimes been attracted by the prospect that science will reveal that the attributes which provoke distaste, disdain, or disgust in others have a genetic basis, believing that public knowledge of this putative fact would inspire greater understanding and sympathy. The hope is probably optimistic, but it does testify to a widely shared attitude that it is wrong to blame people for conditions that they cannot help.

Why, then, should we fear that genetic disclosures will lead to social discrimination? Recognizing someone as carrying genes that predispose him or her to a disease—colon cancer, Huntington's disease, or diabetes, for example—should spark sympathy, not dislike, contempt, or revulsion. Yet sympathy, or "special consideration," may intensify the shadow already cast by the disease; the kindness of strangers is sometimes an additional burden. As we learn of the ills that threaten friends and acquaintances, we may also have to learn more social delicacy, so that they are not made into premature invalids. Although there will probably be genuine challenges here, they are surely not insuperable. Far more worrying is the thought that existing forms of discrimination will provide the basis for expanded or intensified ways of ostracizing people. Perhaps a genetic susceptibility to disease will be associated with some social trait that is now disdained; perhaps beliefs that certain tendencies to action now viewed with scorn or repugnance are caused by particular genes will fix a stigma on those who bear the pertinent alleles, irrespective of what they actually do.

The most likely way in which people bearing disease genes will become vulnerable to a social stigma is for them to suffer the more concrete forms of genetic discrimination. If people at high risk for certain diseases become uninsurable and unemployable, then society will indeed contain a new class with many of the attributes that cause stigmatization: These people will be poor, they will not work, they will be deprived of other opportunities, they may come to have a higher crime rate. Just as failure to solve the problems of insurance coverage echoes in the context of employment decisions, so too inability to prevent either of the tangible forms of genetic discrimination will produce new pariahs.

Beliefs about genes implicated in the springs of human behavior might also fuel discrimination. For much of the twentieth century, the lay public in democratic societies has been periodically titillated with claims that characteristics of enduring fascination—tendencies to violence, intelligence, sexual preference—have a genetic basis. Little imagination is required to foresee the possibility that future acceptance of some hypothesis about the genotype responsible in one of these cases will stamp people with a derogatory label before they have had a chance to act in the offensive way, possibly even when they are barely beyond the cradle. Recalling the discussion of the likely future of molecular investigations and anticipating a more extensive discussion of the role of genes in human behavior, we should note that the distinctive talent of molecular genetics is to identify alleles that have large effects on the phenotype and that are similar (more exactly, *homologous*, similar by descent) to alleles expressed in nonhuman organisms. If (and it is a big "if") the hedges and qualifications that experts typically supply are firmly transmitted to public consciousness, the temptation to think that we can fix firm labels on people, marking them with traits that both repel and fascinate, should evaporate.

Unfortunately, as is clear from recent discussions of a biological basis for sex roles, sexual orientation, criminality, and sex differences in intelligence, many people have a credulous fondness for stories that oversimplify complex phenomena. The problem has two sources. One could be addressed by increased understanding of the intricacies of genetics and by fostering public discussions that expose flamboyant claims to criticism. Yet even if this were accomplished, even if citizens became models of sober discernment, it would still be important to combat intolerance. Whatever the facts about the causes of, say, homosexuality, discrimination against homosexuals—or against those carrying alleged "gay genes"—would be morally unjustifiable. The Nazi programs in which Jews, Gypsies, and homosexuals were sterilized or murdered would not have been vindicated even if the absurd genetic doctrines of "race biology" had been correct.

Potential problems of social stigmatization can be faced more pragmatically, without hoping for great improvements in general understanding of scientific claims or an atmosphere of increased tolerance and respect. Provided that insurers and employers are restrained from using genetic information in the ways suggested above, there will be less incentive for people to acquire knowledge of others' genotypes and the task of protecting sensitive details about individuals may become easier. Rumor and innuendo may continue to flourish, as they do now, but the vision of inquisitive hackers ransacking databases for genetic profiles need not haunt our dreams.

Even without disclosure of the genotypes of particular people, accumulated statistics about the distribution of various alleles may reinforce old prejudices. Three episodes from the recent history of genetics, physiology, and institutional policy within the United States are worth remembering. One, already considered, the exclusion of those with sickle-cell trait from the Air Force Academy, denied African Americans access to coveted opportunities for advancement. A second, the institution of a program in sickle-cell testing, served mainly to label American blacks, and was a botched contrast to the successful Tay-Sachs program launched in the North American Jewish community. Finally, a large chemical company, American Cyanamid, invoked molecular biology to justify barring female employees of childbearing age from working with certain toxic substances, ignoring the fact that the chemicals in question would have mutagenic effects on sperm as well as eggs. Antecedent judgments about the characteristics, abilities, and proper roles of groups that have historically suffered discrimination—such as blacks and women—easily insinuate themselves into the processes of selecting and applying genetic information.

Systematic knowledge of the human genome could correct prejudices. We know already that different ethnic and racial groups are particularly susceptible to different genetic diseases. Whites of Northern European extraction are at greatest risk for cystic fibrosis; people of African descent have the highest frequency of the sickle-cell allele; the Tay-Sachs allele is most prevalent among Ashkenazi Jews; and people from the countries ringing the Mediterranean are most likely to carry the gene for beta-thalassemia. Perhaps future research will show that the catalog of genetic disabilities is distributed across racial and ethnic groups in roughly equal fashion, engendering conviction that, while there are genetic differences, taken as a whole, all groups are "genetically equal" (or, perhaps, equally defective). The fundamental moral point is independent of any such discoveries. *However* the genetic lottery has distributed fortunate or unfortunate holdings should not affect attitudes toward individuals or groups: It would be quite wrong to suppose that *inequality* in the distribution would represent grades of superiority and inferiority among races.

Increased knowledge about human genotypes poses tangible threats of discrimination, which can be met by societies committed to protecting the genetically unlucky. Where universal health care is assured, where genetic information does not block people from securing their futures, where genetic risks do not bar the way to employment, the institutions of society will not add to genetic misfortune. If these steps are taken, it will be possible for our descendants to avoid creating a new class of social pariahs. They will still have to be vigilant, however, recognizing that old prejudices may be cloaked in apparently scientific claims about genotypes. Casual claims about genes associated with unpopular forms of behavior may be accepted uncritically, reinforcing stereotypes and rationalizing discrimination. Indeed, as we learn more about genetic variations, our societies will have to cultivate robust forms of tolerance, appreciating difference instead of insisting on narrow ideals of human worth. Even if there are no new pariahs, constant effort will be needed to ensure that the advances in knowledge do not yield new injustices, new miseries for yesterday's outcasts.

7

Studies in Scarlet

Except for *monozygotic* ("identical") twins, each person possesses a unique DNA sequence. Setting identical twins aside, it would be possible to track people by following their DNA. More realistically, knowledge of DNA sequences can prove useful in identification projects: reuniting families torn apart by war or by the actions of repressive regimes, identifying corpses, checking paternity, and most commonly, investigating and prosecuting crimes. Forensic uses of DNA technology inspire great hopes and arouse considerable controversy.

In 1983, in a small village in central England, a teenage girl was raped and murdered; almost three years later, in a village nearby, another young girl died as the victim of a brutal sexual assault. Frustrated in traditional lines of investigation, the Leicestershire constabulary enlisted the aid of a geneticist at Leicester University, Alex Jeffreys, who was already beginning to earn international fame for his techniques of DNA analysis. Deciding that the murderer was a resident of or a worker in one of three villages, they requested that the young men from the area give a sample of their blood. (Because of the high sperm count in the semen from the sample in the first murder, it was possible to fix an upper limit on the age of the criminal.) With cooperation from the local inhabitants, blood was drawn from over three thousand men, and after several months of collection and analysis, the police found their man. He was not among those who had given. Colin Pitchfork, a local baker, had persuaded one of his fellow workers to donate a sample in his name, and once the police discovered this, they were easily able to establish Pitchfork's guilt.

Pitchfork's respect for the power of science undid him. Had he committed his crimes on a different continent a few years later, he might have been better advised to give the sample and focus his defense on the inconclusiveness of DNA evidence. In the late 1980s prosecutors in several American states claimed to show beyond reasonable doubt that the suspect had left blood or semen at the scene of the crime. Matching DNA from suspect and crime scene, they flourished truly awe-inspiring figures: The probability of a match occurring by chance was set at one in a million (or a billion). Closer scrutiny of the calculations, and discovery that companies performing DNA analyses (most notably Lifecodes Corporation) were not models of scientific or clinical rigor, provoked a backlash. Supreme courts in some states declared DNA evidence inadmissible.

Yet the allure of forensic uses of DNA is obvious. Violent criminals find it hard to avoid leaving mementoes of themselves at the scene of the crime and, properly applied, DNA technology would seem to be a powerful tool both for catching them and convicting them—as well as for exonerating those wrongly accused. Alex Jeffreys, the geneticist who pioneered applications of DNA technology in forensics, has expressed the hope that genetic "fingerprinting" will have an enormous impact on rates of violent crime—presumably, in part, by confining perpetrators and, in part, by deterrence. Is this the characteristic optimism with which brilliant scientists foresee the future of their intellectual progeny? Or can we make DNA analysis into a precise and valuable forensic tool, something that will work reliably without trampling the rights of defendants or burying the innocent under a barrage of misleading numbers?

• • •

At present, the typical method of using DNA analysis in forensic contexts involves comparing bodily materials (blood, semen, hair follicles) left at the scene of a crime with a sample (blood or a cheek swab) taken from a suspect. Purifying DNA from the crime scene sample is a delicate enterprise and, quite commonly, analysts are able to extract only a small amount of DNA containing only parts of the criminal's genome. The next step is to look at regions of the DNA that are known to be highly variable (such as RFLPs or VNTRs; recall that these are regions where differences in the nucleotide composition of the DNA can be detected). Often the DNA is amplified using the polymerase chain reaction (PCR; see Figure 7.1), an experimental technique used to obtain a collection of fragments that are then separated on a gel. Sometimes, when the sample is sufficiently large, it is chopped up with restriction enzymes, yielding fragments for gel electrophoretic separation; the crucial pieces of DNA are then picked out using radioactive probes. The product of these intricate procedures is a *genetic profile* of the criminal, a specification of the characteristics of the DNA he (or she) bears at loci that prove diagnostically valuable. (Despite the popularity of the expression, I shall avoid calling this a "genetic fingerprint" since, as we shall see, whether genetic profiles individuate people in just the ways fingerprints do is a major point of dispute.)

Analyzing the sample from the suspect is far easier, since it is not hard to obtain as much DNA as would prove useful—even a small amount of freshly drawn blood provides the basis for many more tests than can typically be undertaken with the degraded, often contaminated, samples taken from the scene of the crime. If we could generate the complete sequence found in the crime scene DNA and compare this with the complete sequence from the suspect, there would be no uncertainties: Monozygotic twins aside, two people are expected to differ by about three million base pairs. Complete sequencing is impracticable, however, both because of the scale of the project of sequencing any mammalian genome and because forensic analysts have to grub for small amounts of DNA where they can (un-

Original two-stranded DNA

Separate strands and add primers:

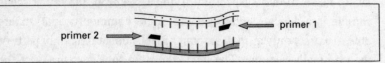

primer 1

primer 2

Extend complementary strands from primers:

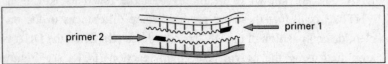

primer 1

primer 2

Separate strands and add primers:

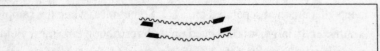

Extend complementary strands from primers:

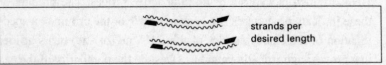

strands per desired length

Repeat to obtain more strands of desired length. (At each stage there will be some strands of the wrong lengths, but these will soon be overwhelmed by a vast number of strands of the desired length.)

FIGURE 7.1. *The Polymerase Chain Reaction (PCR)*

PCR is a method for making many copies of a segment of DNA, provided enough is known about the regions flanking the segment. At the top is a piece of double-stranded DNA, from which a segment is to be copied. The strands are separated, and appropriate "primers" are added, binding to short pieces of the single strands that surround the segment. When bases and appropriate other molecules are added, complementary strands are extended from the primers: Many of these strands will be much longer than the ones desired. The strands are again separated, and more primers are added. This time some of the complementary sequences formed will have just the right length, because they will effectively have primers at both ends. As we proceed, the proportion of segments of just the right length will become larger and larger, until, after the process has been repeated twenty times, virtually all the DNA in the mix will be a copy of the exact segment we wanted to copy.

less the samples are reasonably fresh, there is a significant chance that they will contain only parts of the total genomic DNA). So it is necessary to choose bits and pieces for analysis, hoping that they will be the regions where individual differences show up.

Submicroscopic creatures touring our chromosomes would probably discover that some areas—such as those responsible for directing the formation of vital proteins—are virtually invariant from person to person. Coding regions (segments of DNA that are transcribed and that direct the formation of proteins) usually show only limited variety, because changes in sequence, apart from those that do not affect the amino acid encoded, easily result in proteins unable to perform essential tasks. DNA analysts want to avoid these regions, focusing instead on places that are least under the control of natural selection, areas of no functional significance such as the expanses of "junk" DNA that separate our genes. Molecular geneticists already know some areas where numerous different sequences of A's, C's, G's, and T's can be found in members of our species, and systematic study of the human genome is likely to disclose further variable regions of nonfunctional DNA, which can serve as ideal "meaningless markers." The optimal result, of course, would be to discover a region in which everyone's sequence is different from that of everyone else.

Genetic profiles are presently constructed by looking at the DNA in several highly variable regions. Examining a single stretch will not normally suffice, for although there may be many alleles at a RFLP locus, there are not enough to differentiate people. When two loci are considered, the probability of a chance match is reduced. Two people might have the same combination of alleles at a *single* highly variable locus, but there is only a very small probability that the alleles will be the same at *two* highly variable loci, a truly minute probability that they will be shared at *three* highly variable loci, and so on. Even though the population of potential perpetrators may be quite large, concentrating on multiple loci might yield a very low probability that anyone else in the population, besides the suspect, has the same combination of alleles found in the DNA left at the scene of the crime.

For the moment, let us suppose that there are no technical problems that cause mistakes in determining the characteristics of the DNA in either the crime sample or the sample taken from the suspect. The first issue to address concerns what we are entitled to infer from a perfect match. A *mis*match is unambiguous. When the samples show different combinations of alleles, the suspect can definitely be excluded from further investigation. Indeed, in recent years several people who had been wrongly convicted have won release from jail because DNA tests showed clearly that they were not guilty of the crimes for which they were being punished.

On the other hand, if the genetic profiles coincide, we can conclude that the sample at the crime scene might have come from the suspect—but, of course, it might have come from someone else who happens to have the same profile. If we knew the profile of everybody in the population, we could determine if the latter possibility is genuine. Without that knowledge, analysts have to estimate the chances.

Initially, the problem of showing that the match is significant appears perfectly straightforward. To share the sequence at one locus with the criminal is unfortunate; to share the sequences at several loci begins to look like carelessness. More precisely, we ought to be able to use statistics on the relative frequencies of alleles at the loci we analyze to calculate the probability that the profiles might match fortuitously. Suppose there is a chance of 0.01 of a match at the first locus, a chance of 0.005 of a match at the second locus, and a chance of 0.002 of a match at the third locus. Then it is quite natural to think that the probability of a match at all three loci is the product of the individual chances, 0.0000001, or 1 in 10,000,000. Quite natural, perhaps, but potentially wildly incorrect. Multiplying probabilities works beautifully when we are thinking about the chance of obtaining three heads on three tosses of a coin, or contemplating the probability of winning three horse races, because the events are independent. When independence fails, unless we have further statistical information, the most we can say is that the probability of all three matches is less than or equal to the lowest of the probabilities of sharing at the individual loci. If everyone with the rare combination of alleles at the third locus has exactly the same profile at the other two loci, then the probability of a total match is not 1 in 10,000,000, but only 1 in 500.

A comparison with more familiar features that distinguish people from one another may highlight the point. We know that observable physical characteristics come in clusters. Imagine that an eyewitness to the crime testifies that the criminal has red hair, green eyes, and freckles. The suspect has all three traits—and wears the map of Ireland on his face into the bargain. The prosecutor points out that only 1 person in 100 has red hair, only 2 in 100 have green eyes, and only 1 in 20 has freckles. Multiplying the probabilities, he argues that there is a chance of only 1 in 100,000 that someone in the general population has all three characteristics. The calculation is clearly wrong: The probability of having green eyes if you have red hair is significantly larger than the general probability of having green eyes, and the probability of having freckles if you have red hair and green eyes is much greater than the probability of having freckles. People do not mate at random, and characteristics that are individually quite rare associate in particular ethnic or racial groups, belying the tactic of multiplying the probabilities for all three characteristics individually. The Irishman in the dock may be guilty, but a significant number of his countrymen would fit the witness's description.

How, then, should we calculate the chance that someone other than the suspect has a genetic profile matching that obtained from the scene of the crime? Two obvious strategies come to mind: We can try to choose to analyze parts of the genome that are quite unconnected with one another, so that the sequence in one place would not be expected to be correlated with any particular sequence at another place, or we can attempt to find a different way of computing the probability of fortuitous match. At first sight, there seems to be an easy way of carrying out the first strategy.

Suppose we considered stretches of DNA from different chromosomes. Recall that one of each pair of chromosomes is passed on in the formation of sperm and ova: You have two copies of chromosome 1 (call them 1A and 1B) and two copies of chromosome 2 (2A and 2B), and whether one of your children receives 2A or 2B is not affected by which of 1A or 1B you passed on. Enough generations of random mating will mix up genetic material from different chromosomes to produce at least approximate independence—so that there will be a very low degree of correlation of sequences across chromosomes.

However, when people have a tendency to marry and have children with people like themselves, all bets are off. Trouble arises, in particular, if those who marry are homozygous at the locus under investigation, that is when they carry two copies of the same allele at that locus. If Boris and Doris are both homozygous for particular sequences of DNA on two different chromosome pairs, say chromosomes 1 and 2, both carrying the same sequences—both have two copies of the sequence S at the locus analyzed from chromosome 1, and both have two copies of the sequence S^* at the locus analyzed from chromosome 2—then their children will (almost inevitably) have exactly the same sequences. So Chloris and Maurice will show just the association between a region on chromosome 1 and a region on chromosome 2 that their parents did—they, too, will have two copies of S at the analyzed locus on chromosome 1 and two copies of S^* at the analyzed locus on chromosome 2. In fact, it does not matter if the people involved carry exactly the same sequences. So long as the differences are indiscernible by the methods employed in forensic analysis, the different parts of the genetic profile will not be independent.

Inbreeding causes trouble. The well-known tendency of people to choose partners from their own ethnic group doubtless produces subpopulations of our species that are inbred to some extent. What are the consequences for the independence of the loci considered in forensic analysis? Nobody knows. They are probably different for different subpopulations, but the statistical tests for showing that the distribution of genes in human populations approximate the results of random mating are extremely delicate—so delicate, in fact, that population geneticists can be recruited for public defense of different conclusions.

The alternative strategy is to find an accurate way to compute the chance that genetic profiles are identical, without assuming that the constituent loci are independent. Elementary principles of probability yield a formula for the joint occurrence of three events: The chance of all three is the probability of the first multiplied by the probability of the second, *if the first occurs*, multiplied by the probability of the third, *if the first and second occur.* So it would have been correct to calculate the probability of a red-haired, green-eyed, freckled man as the product of three factors: The probability of having red hair, the probability of having green eyes given red hair, and the probability of having freckles given red hair and green eyes. If forensic analysts knew the probability of each combination of alleles at any locus given any of the possible combinations at other loci, then they would be able to compute the probability of fortuitous match precisely. Unfortunately, the task of assembling all the pertinent probabilities would be truly Herculean. The stretches of DNA needing investigation are precisely those with high variation, and so the number of probabilities needed is the product of several large numbers.

Perhaps the problem can be made more tractable by not insisting on an *exact* figure? Recognizing that the frequency of an allele in a subpopulation depends on which subpopulation we pick (as, for example, some disease alleles, like the CF allele, are more common among people of Northern European extraction), we could sample populations of people with different ethnic backgrounds, cautiously choosing the highest frequency at each locus to estimate the probability that two different people will share the same profile. Provided that the number of groups examined was not too large, the project of gathering the statistics would become manageable. The idea would then be to proceed as if the loci were independent, multiplying the probabilities but compensating by choosing conservative (high) values for the individual probabilities. Unfortunately, there is no reason to think that this approach will provide anything close to the truth. The exact figure for the probability of matching profiles is obtained by multiplying *conditional* probabilities, the probability that a particular combination of alleles will be present at one locus *if* certain others are found elsewhere. Those figures might be approximated by the highest frequency with which the combination of alleles is found in various groups of people—but there is no reason for this to be so. Blindly compensating for possible failures of independence will not provide a reliable estimate of the true chance.

Imagine that a new prosecutor continues the case against the unlucky Irishman. Suppose that: The population of people in a position to commit the crime numbers 5,000; and, of these, 1,000 are Irish; and 1 in every 100 of the latter has red hair, green eyes, and freckles. So there are ten Irish people who fit the witness's description. The true frequency of the combination in the population of possible perpetrators is thus 10 in 5,000, i.e, 1 in 500. The new prosecutor compiles statistics for five ethnic subgroups of this population, including the Irish. He discovers that the frequency of each of the traits is highest among the Irish: 1 in 80 has red hair, 1 in 70 has green eyes, and 1 in 20 has freckles. Multiplying these numbers together, he calculates the frequency of the combination of traits as 1 in 112,000. Because the police have established that there are only about 5,000 people who could have committed the crime, the jury is convinced that the possibility of a fortuitous match can be dismissed.

Given the prosecutor's figure for the crucial probability, it turns out that the chance that someone else in the pool of possible perpetrators has the three traits is about 1 in 25. (To calculate this probability it is enough to notice that for each person in the pool of possible perpetrators the probability of having a profile that does *not* match that of the Irishman is 111,999 in 112,000; so the chance that all 5,000 have profiles that do not match the accused's is 111,999/112,000 raised to the power 5,000; that number is around 24/25; so the chance that someone in the pool of 5,000 has a profile that matches is 1/25). If the correct probability, 1 in 500, had been accepted, the chance that the pool of potential perpetrators contains at least one other person with red hair, green eyes, and freckles is over 9,999 in 10,000 (the chance is calculated in just the same way, namely by first computing the probability that *nobody* in the pool will have a matching profile).

The tight association of the traits is hidden from view because it occurs in a very small proportion of one of the classes on which statistics are compiled. Faced with analogous problems, forensic analysts might try to correct the deficiency by studying more finely differentiated groups. Not only would this vastly increase their labor, but they would remain vulnerable to the suggestion that an association of traits may be concealed within one of the populations they have studied. At present, the case for the prosecution never rests on any fine-grained analyses. Prosecutors proceed on the assumptions that the populations relevant to the crime are well mixed and that the failure of independence exhibited in the hypothetical example of the last paragraph will not occur. Many population geneticists would contend that they are quite justified to do so, but others believe that until the actual statistics have been collected we cannot be confident that apparently plausible reasoning has generated the correct estimates of probability.

All the approaches so far have attempted to work out an overall probability that genetic profiles will coincide from knowledge of the frequencies with which constituents of the profile (the combinations of alleles at various loci) occur. A more direct way, when police can specify a pool of potential suspects, is to investigate whether the overall genetic profile is found elsewhere in that pool. If Colin Pitchfork had been bold enough to give blood at an early stage of the investigation, the Leicestershire police could have guarded against the possibility of a fortuitous match by continuing to sample the rest of the population. (So, perhaps, even with the doubts that have come to surround the use of DNA evidence, but without his confession, Pitchfork might have been justly convicted.)

The investigation might go like this. After discovering a match between the suspect's DNA and that from the scene of the crime, analysts take samples from a significant proportion of the people who were in a position to commit the crime. Finding that nobody in this group has a matching genetic profile, they conclude that the probability of fortuitous coincidence is low, and that the material left at the crime scene came from the suspect. The more thorough the sampling, the more warranted is their conclusion. Since none of those sampled has the crucial genetic profile, the probability that an *unsampled,* member of the pool of potential suspects has the profile is less than the reciprocal of the sample size. So if 1,000 people have been sampled, the chance is less than 1 in 1,000. This means that the probability that an unsampled person doesn't have the profile is greater than 999 in 1,000. The probability that the *entire* unsampled residue of the pool contains nobody with the profile is this quantity (0.999) raised to a power, the power being the number of unsampled people. So, if 100 people remain uninvestigated, the chance that all of them lack the profile is $(0.999)^{100}$, about 9 in 10 (0.9). Plainly, the closer analysts are able to come to investigating the whole pool, the stronger the inference. In the limit, as they round up *all* the potential suspects, it becomes as definite as their knowledge of who would have been able to commit the crime.

In practice it will usually be difficult to ensure that all those who could have committed the crime provide materials for analysis. The case for the prosecution will have to rest on a probability calculated by sampling from the pool of potential perpetrators. Thoroughness in sampling is not simply a matter of making sure that the unexamined residue is as small as possible—eliminating potentially significant differences between those investigated and those uninvestigated is also crucial. Allowing the supply of blood (or other bodily materials) to be voluntary, as in the Pitchfork inquiry, would thus be counter-productive, since the real criminal would be expected to try to avoid giving blood. Hence, the mere fact that a fortuitous match does not appear among the innocent volunteers provides no basis for extrapolating to the remainder of unexamined potential suspects. Similarly, if a particular ethnic group occurs in appreciable numbers in the pool of potential suspects but is underrepresented among those investigated, there will be grounds for questioning the calculation of probabilities.

On many occasions, forensic scientists could overcome these difficulties if only they could take the first step of circumscribing a manageable pool of potential perpetrators. Sometimes the conditions of the classical murder mystery are approximately satisfied, and detectives can reasonably assume that only a relatively small number of people could have been in the right place at the right time (as, for example, the Leicestershire constabulary concluded, correctly, that only men familiar with the village would have been in the isolated spot where the first murder occurred). Frequently, however, picking out the potential criminals depends on quite substantive assumptions about the way in which the crime occurred. When a violent crime takes place in a large city, narrowing the range of scenarios may be very difficult. Although it seems most likely that the criminal was from the immediate vicinity, one cannot exclude the possibility that he is already thousands of miles away by the time the crime is reported.

Pursuing the direct approach by investigating the frequency of the overall genetic profile among those viewed as having the opportunity to commit the crime requires the prosecution to do significant work in justifying its conception of the potential criminals. DNA evidence holds out the promise that it will replace qualitative considerations with numerical precision. In the end, however, the exact figures gain evidential force only against the background of more traditional kinds of argument, which show why it is right to concentrate on particular populations.

The troubles with which we have been wrestling stem from the fact that, so far, the extent to which genetic profiles individuate people—the extent to which they deserve the name of "genetic fingerprints"—is unknown. Resolving the uncertainty is not impossible, and we shall consider ways of doing so shortly. First, however, it is important to consider some of the other potential troubles with forensic uses of DNA.

Announcing extraordinary figures to show how improbable it is that anyone except the defendant could have left the sample at the scene of the crime may make for high courtroom drama, but—in addition to the dangers of miscalculation we have reviewed, and in addition to the very real chance that jurors will misunderstand the probabilities—the estimates are often pointless. If the defense can argue that there is a significant chance that mistakes have been made in processing or analyzing the samples, it makes no difference whether the probability of fortuitously matching genetic profiles is 1 in 10 or 1 in 10,000,000.

Error can creep in in a number of different ways. Some of them wear their failure on their faces. If a technician forgets to adjust the temperature during a crucial reaction or lets a gel run too long, the blank results will show clearly that something has gone wrong. The problematic mistakes are those that appear to show a match in genetic profiles when there is none. Two simple ways for this to occur involve mislabeling or contamination. If at some early stage of processing, the materials from the crime scene are confused with those of the suspect, one set of materials may be marked with two different labels, so that a match occurs because the same DNA is analyzed twice. Even if the labels are correct, a similar result can be produced if the apparatus used in the analysis is not thoroughly cleaned between the processing of one sample and the processing of the other. Forensic techniques have to be sensitive to minute amounts of DNA—otherwise the poverty of the materials deposited at the crime scene would doom the enterprise—and a tiny trace of the first sample can contaminate the analysis of the second.

Comparing genetic profiles is a matter of deciding if two processes of generating DNA fragments yield pieces of the same size. Analysts determine relative size by seeing how far the fragments migrate through a gel, inspecting the pattern of bands produced on an X-ray film. Because gels vary in density and porosity, it is unrealistic to expect that the same fragment will go exactly the same distance through two different gels. Consequently, it is important to run one or more *control lanes*, using fragments of known size to serve as a ruler for measuring migration distances. In this way, even though the pictures of the bands may not show fragments in precisely the same places, it is possible to recognize that the shifting of relative position has occurred as a result of the idiosyncrasies of the gels.

Reading the pictures is sometimes an art. Not only do interpreters have to cope with the possibility that bands have been shifted, but they must also make judgments about whether the faint splodge on one film really counts as the clear band exhibited on the other. Judgments like these are plainly vulnerable to antecedent expectations. Nineteenth-century craniometers set out to determine the skull sizes typical of different racial groups, believing, quite erroneously, that brain size would measure native intelligence, and they recorded apparently objective results that seemed to show exactly the relationships corresponding to their prior convictions. Stephen Jay Gould has lucidly analyzed how filling the skulls with lead shot, and comparing the weights of the lead, could easily be infected with unconscious biases—and how his efforts to reenact the craniometers' experiments overlooked details that interfered with his own preconceived notions. Delicate experiments are often perturbed by what the experimenters think they know, even in situations where scrupulous researchers are trying to guard against the urgings of background belief. When a defendant's future, possibly his life, is at stake, it is important to reveal the interpretative art lying behind the apparently objective numbers.

In the early days of forensic DNA analysis in the United States, revelations of the art proved embarrassing. In one notorious case,

Lifecodes had trouble deciding on the sex of the person who had contributed the supposed control sample. Since then, with greater awareness of the pitfalls, companies offering DNA analysis have improved their records. However, the teething pains of the technology suggest obvious morals for monitoring future uses.

First, the dangers of confusing samples can be significantly reduced if the materials from the crime scene and those from the suspect are processed by different laboratories. Second, defendants should be sent copies of the original pictures and, if the prosecution decides to claim a match as evidence, the defense should have access to experts accustomed to reading such pictures. Third, and perhaps most important, the performances of laboratories and of analysts should be regularly calibrated. This can easily be done by testing the success rate in determining matches among coded samples: The evaluators know which vials of blood came from the same person and which did not.

The rights of defendants will be properly protected if forensic uses of DNA occur within an institutional framework that requires both independent testing of samples by people who have no stake in securing a conviction and whose performances are constantly assessed, as well as cross-checking of interpretations to reveal potential biases. Economic considerations, and horror of bureaucratic complications, may point in another direction, toward forensic analysis by the local police and their backroom laboratory. Without safeguards against the errors of contamination and eager overinterpretation, DNA analysis may prove cheap—but it will be no bargain for those who stand in the dock.

Reliable knowledge of match or mismatch between genetic profiles can be achieved, but we shall still have to confront the earlier problem of deciding the exact significance of a match. There is a simple way to free ourselves from worries that probabilities have been miscalculated and to increase the power of forensic DNA analysis besides. Any country could try to construct a national DNA database, registering a genetic profile for each person born within its borders and demanding that a profile be provided for everyone (legally) entering. Estimating the chance that someone other than the suspect had a genotype matching the DNA in the crime sample would no longer be necessary: It would suffice to consult the database. Moreover, the database would serve as an instrument of discovery. Abstracting from complications involving monozygotic twins, police officers could search a database to find the unique individual whose DNA matched that found at the scene of the crime.

The vision of a national database appalls many citizens, conjuring up the prospect of Big Brother scrutinizing the private actions of innocent people. Would a national database constitute an unwarrantable invasion of privacy? That depends very much on what it would include, how it would be generated, and how it would be employed.

Consider first how genetic profiles might be constructed. Systematic exploration of human genomes is likely to provide clues about regions in which there is greatest variation in sequence. Choosing from these regions segments of DNA susceptible to reliable analysis, it should be possible to discover loci at which the most common combination of alleles occurs at a frequency of less than 1 in 10. If twelve such loci were chosen, then, provided that the alleles assorted independently, there would be more than a trillion twelve-locus genotypes, more than enough to provide one for everybody on the planet. Of course, breakdowns in independence might have the consequence that only a small fraction of these were represented among the citizens of the nation (or nations) aiming at universal typing, but if it were discovered that twelve loci could not assure individuating genetic profiles (for everyone except monozygotic twins), the remedy would be to add further loci to disambiguate the troublesome cases.

What exactly would be represented in the database? For the sake of simplicity, we can assume that the loci chosen involve a variable number of short repeats and that twelve prove sufficient. Each citizen can be represented by a short sequence of numbers. Thus the code for one person might be:

57 26 141 12 11 49 61 82 14 35 101 19.

The numbers would reveal how many occurrences of the unit the person had at each locus. So, in the example, the person has 57 copies at the first locus, 26 copies at the second, and so on. People would be registered by entering the appropriate sequences of numbers with their names attached. The samples they had supplied could be, and should be, destroyed, so that no further information can be obtained from them. All that would remain would be a sequence of numbers disclosing no exciting properties. For the fact that someone has 57 occurrences where another has 29 has no importance whatsoever, since these highly repetitive regions are remarkable for their functional insignificance.

How would samples from an entire national population be obtained? Newborn infants are sitting targets for the nurse's needle; immigrants and short-term visitors could be asked to donate blood or a cheek swab at the border (or perhaps to produce evidence of past typing). Adults who had not been typed as children might be required to attend medical centers at which samples would be taken at no charge, and the state could enforce the demand by insisting on proof of typing before acquiring a driving license, receiving unemployment benefits, entering a hospital, and so forth. The most difficult people to include are those loosely connected with the social system, but even here dedicated effort should be able to supply profiles for a large majority. Typing virtually all the citizens of any of the affluent democracies is a realistic possibility, and if only a handful of people are excluded, evidence showing coincidence of genetic profiles retains its force. For if the genetic profiles of millions of people from all the ethnic backgrounds represented in the country show coincidence in genetic profiles only in the case of monozygotic twins, protests invoking the bare possibility that someone who has slipped through the net shares the pertinent DNA sequences with the accused are bound to ring hollow.

How would the database be used? After a crime has been committed, any human DNA obtained from the scene would be amplified at the crucial loci and the number of repeats at each locus would be determined. A sequence of twelve numbers would be entered in the database for a possible match. (If the sample from the scene was sufficiently degraded, it might not be possible to analyze all the loci, so that only part of the sequence would be entered.) If there were a match to the profile of a single individual in the population, the police would use standard methods to find the person and would be permitted by law to demand a further sample. The new sample would be sent for analysis to a different laboratory, one that had examined neither the crime scene sample nor the sample used in the original typing. If the match were sustained, the new sample would be retained as potential evidence in any trial.

The system I have described endeavors to meet the most important concerns about forensic uses of DNA. The database can be employed to exonerate, to convict, or to discover important facts about the criminal. Destruction of the samples restricts the information on record to a sequence of meaningless numbers. Mistakes in analyzing DNA will probably always occur at low frequency, but requiring a second sample in the case of a match prevents suspects from becoming victims of those errors, unless exactly the same mistake is committed twice. Insisting on analysis by independent laboratories makes the chance of that occurring almost infinitesimal.

Worries may linger. Wouldn't a national database make it possible for demented bureaucrats to track the movements of citizens? Bureaucrats engaged in any such project would be truly demented, for the work involved in finding and analyzing the DNA would be grossly disproportionate to the value of the information acquired. Couldn't corrupt police officers substitute a blood sample from an innocent suspect for the crime scene sample, thus framing someone they want to convict? Indeed they could, but there is nothing new here. Corrupt policemen already have ample means of incriminating the innocent. They can plant any number of things that will connect their victim to the crime. Even without a national database, they can claim to have found a blood sample, taken from the scapegoat at the scene, and after sending it for analysis, announce that the genetic profile matches that of a fresh sample obtained when the suspect is brought in for questioning. The worry ultimately rests on a schizophrenic attitude toward powerful forensic methods: Most of us want the law to have effective ways of finding and convicting the guilty; most of us fear that effective methods could be abused. The obvious remedy, easier to state in principle than to achieve in practice, is to erect safeguards against official misconduct, not to resign ourselves to corruption and allow only second-rate techniques.

To allay fears of an Orwellian future, it is helpful to recall the Pitchfork inquiry. Without the voluntary contributions of blood from almost all the young men in three Leicestershire villages, the police strategy would have fizzled. No doubt many of those who gave experienced considerable social pressure, but the nudgings of wives, girlfriends, and parents surely symbolized a commitment that the innocent must work together to expose the guilty. Moved by two terrible crimes, the community resolved on the joint disclosure of meaningless information. Instead of thinking of a national database as baring citizens to Big Brother, we might better view it as the cooperation of the innocent to distinguish, and so protect, themselves from those who perpetrate violent crimes. Insignificant disclosures by sufficiently many people would force the agents of aggression out of the shadows, promoting the security of their potential victims.

Yet do we need to record a genetic profile for everybody? Wouldn't it be enough to compile a partial database, including special segments of the population—convicted criminals and members of the armed forces (to ascertain the identities of those who die in battle)? Perhaps we would discover that in this large population a particular choice of loci for constructing genetic profiles assigned different profiles to different people (with, as always, the exception of monozygotic twins). Given a match, we could now be confident that it signaled the presence of the suspect at the scene of the crime. Moreover, since most crimes are committed by recidivists, the extra work and expense of extending the database to the entire population would bring no significant returns.

Seductive though it is, this argument threatens to give new meaning to the idea of rounding up the usual suspects. In countries (such as the United States) in which the criminal population and the armed forces do not adequately represent the range of ethnic backgrounds present in the nation, it would be unwise to extrapolate from the fact that genetic profiles individuate people in these particular groups. The lore about recidivism could easily become a self-fulfilling prophecy. Criminals are identified in terms of convictions; convictions would be generated, in large measure, by comparing a crime scene sample to a database, largely made up of previous criminals; recidivists are caught more frequently because they are the prime targets of inquiry.

No doubt universal DNA typing would be expensive. At likely costs of a hundred dollars per person, constructing genetic profiles for millions of people would cost hundreds of millions of dollars. Of course, the costs must be measured against expenses that are currently undertaken in criminal investigations. Yet in the end, economic considerations should not take precedence over the concerns of justice. The most decisive way to show that a genetic profile picks out a unique person would be to have everybody on record. A national database would also best accord with the demands of justice, refusing to allow some people to be more vulnerable to the law than others. The committed communities of rural Leicestershire offer an example to us all.

So far I have taken for granted Alex Jeffreys's hopeful vision of the future of forensic uses of DNA analysis, assuming that the techniques, if properly applied, would facilitate the detection and conviction of violent criminals. The assumption deserves scrutiny. How often will it be possible to find incriminating traces at the scene of the crime? What, if anything, can criminals do to evade detection?

In most cases of rape, a vaginal swab from the victim will yield sperm carrying DNA from a single source. Gang rapes (or rapes involving several rapists) are more complex, because of the difficulties of analyzing DNA from many people, but the problems can sometimes be overcome. With assaults and murders, police have to depend on blood spilled at the scene of the crime or on cells from the criminal that may be found on the victim (most notably, blood and skin under the victim's fingernails).

Placing someone at the scene of a crime does not show that the person did the deed, or even that any crime has been committed. Rapes usually enable police to establish a direct link between violator and victim, for the rapist's sperm is typically found in the victim's vagina, but prosecutors still face the task of showing that the act was not consensual. By contrast, in most cases of murder and assault, there is no doubt that a crime has been committed, and the chief difficulty lies in showing that the organic samples signal participation in the criminal act. If blood or skin are found under the victim's fingernails, or if it is clear that samples are present at the crime scene because of the victim's efforts at retaliation, the connection can be made. However, when hairs from the victim's spouse are found on the victim's clothing, the match hardly compels the conclusion that the spouse committed the crime.

Victims can make it easier for police to obtain samples and to establish the crucial connections—if they act to ensure that blood, skin, or semen end up in incriminating places. However, violent criminals (and their victims) are likely to change their tactics in the wake of improved forensic techniques. Fingerprint technology prompted burglars to wear gloves. In similar fashion, we can expect rapists, murderers, and assailants to amend their ways. Their actions make it hard to avoid leaving some incriminating evidence. Those who plan can try to block the obvious ways in which cells are shed: Prescient rapists can use condoms, murderers can cover themselves to prevent their victims from clawing them. A more effective approach would be to take advantage of the fact that samples of other people's DNA—unlike other people's fingerprints—are portable. The canny criminal of the future may carry liberal supplies of other people's hair, blood, and semen to be appropriately distributed at the scene of the crime. If these scenarios seem unlikely, it is worth noting that there are already anecdotes of New York prostitutes starting a market in used condoms, apparently so that the buyers can use the contents to divert suspicion.

Many crimes are not premeditated, and forensic uses of DNA may well help catch their perpetrators. Moreover, for all the criminal's forethought, a determined victim may be able to ensure that DNA from the assailant is left behind and that it can be identified as the criminal's. Victims who survive can guide detectives to appropriate places. Even the dead can speak to avenge what has been done to them, if clawing and biting forces samples to be left on teeth or nails. Of course, resolute action may escalate the arms race, intensifying what is already a horrifying encounter, but potential victims do have some ability to threaten. Gauging the extent of their power is a snap judgment that will sometimes determine their fate.

Jeffreys's forecast about the detection of virtually all violent crimes through forensic DNA analysis is, unfortunately, too optimistic. Successful uses of the new technology would change the relations between criminal and victim in unpredictable ways. Doubtless adding a powerful new weapon to the arsenal of the law will increase rates of detection and conviction, removing dangerous people from circulation—and, we should remember, it will also prove valuable in clearing the innocent. Yet it would be naive to think that criminals will remain passive in the face of the new techniques, or that the prospect of inevitable capture and incarceration will stay their hand. Helpful though forensic DNA analysis may be, it is not a panacea.

INTERLUDE

The Specters That Won't Go Away

When the plumbing breaks down, when the new computer only responds with imperious messages and incomprehensible questions, when the car emits peculiar noises, the inexpert—like me—turn to those who know how to fix the trouble. The solutions may not be elegant, for, as the wonder workers often point out proudly, the proper parts are no longer available, but their practical expertise enables them to find substitute ways of getting the job done. Grateful consumers typically care less about the details of the repair than about the prospects of smooth functioning in the future. So long as the water continues to run through pipes that do not leak, so long as the computer behaves like an ally and not a hostile interrogator, so long as the car purrs, we are satisfied.

Readers of previous chapters may view me as having approached the promises and pitfalls of molecular genetics much in the spirit of Mr. Fixit. Brilliant science will give our descendants the power to do good, increasing their ability to diagnose and predict disease, providing new ways to prevent or treat conditions that now afflict millions, fashioning weapons for battling violent crime. As we have seen, unthinking applications would bring hardship as well as benefit. Blindly inserting new technologies into the existing institutions of democratic societies—allowing medical testing to respond to the commercial interests of biotechnology companies, permitting insurers and employers to make free use of the new information, not facing the problems that attend forensic uses of DNA—would cause pain and injustice. We face definite, practical problems: How do we amend current institutions to obtain the benefits without imposing new burdens? How do we make genetic testing work for everyone? How do we avoid genetic discrimination? How do we use DNA evidence justly? Tackling these problems one by one, I have suggested some tentative lines of solution that are often incomplete and inelegant, constrained by the materials at hand—the existing procedures of affluent democracies.

The story I have told runs as follows. The first fruits of the advances in human molecular genetics will be increased powers to test. Genetic tests will enhance the ability of our descendants to diagnose and to disambiguate existing diagnoses. Sometimes, predictive tests will inform patients that they are at high risk for future diseases and, in an unpredictable proportion of these cases, those diagnosed will be able to act to reduce the chances; sometimes, couples will receive the sad news that a fetus bears a disabling genetic condition and, by terminating the pregnancy, they will prevent a life of suffering or early degeneration. Coming decades will also bring extensive knowledge of human genes and their functions as we pry apart those basic mechanisms we share with our evolutionary relatives. To an extent we cannot yet foresee, that knowledge will be translated into action, and our successors will patch together a collection of molecular therapies that may, we hope, transform the treatment of major human diseases.

From the very beginning, it will be crucial to prepare the right social settings for the new applications of molecular medicine. Genetic tests must be soberly appraised and introduced with attention to the needs of the tested. Tomorrow's medicine should not be confined to the privileged, and genetic counselors will have to find ways of enabling all kinds of people to use the new methods to advance their individual interests. In a world where medical records routinely contain genetic information, the patient will need to be protected against institutions or individuals who would use that information to his or her disadvantage. An important part of that protection would be a system of guaranteed health care for all, and there will also need to be regulations on the uses of genetic test results in setting other insurance premiums, as well as in hiring and firing. If workers are to be properly protected against the accidents of their biological inheritance, then, I have suggested, our societies must be committed to creating jobs that will release those trapped between choosing idle health or risky work. Safeguards will also be required in forensic uses of DNA technology—in forging a powerful legal weapon, the courts must be scrupulous in recognizing the possibilities of faulty reasoning that would convict the innocent. Finally, beyond the enterprises of reshaping the institutions of medicine and the law, our descendants should be committed to fostering both greater tolerance and a more critical attitude to claims about genetic determination, lest they write new chapters in the long history of prejudice and discrimination.

And that, one might think, would be that. Here is the promise. Here are the problems. Here are some potential ways of addressing those problems. *Ipse dixit* Mr. Fixit. The issues already discussed are genuine, they are concrete, they are immediate, and they are important: An essential part of the task of preparing for the applications of molecular genetics is to find solutions. Yet even if the concrete difficulties have been resolved in piecemeal fashion, fears about the future would linger. I believe that adequate discussion of the implications of molecular biology for the human future must go further, that it is important to try to exorcise persistent ghosts.

Some specters have already poked their way into previous pages. Reflections on genetic testing have a way of leading to thoughts about prenatal testing—and so to questions of the morality of abortion and of choosing people. As we undertake widespread genetic testing and gene replacement therapy, are we committing ourselves to revive eugenics? How do we decide where treating disease stops and an unsavory creation of modified people begins? Worries about the more indefinite forms of discrimination provoked fears that DNA may be, or may be perceived to be, destiny. To what extent are our characteristics, including our tendencies to act in the ways that matter most to us, determined by our genes?

Departing repairmen often leave sneaking worries as well as gratitude in their wake. Loose ends and odd noises arouse concern. Readers of previous chapters may well wonder if the solutions I have outlined can be sustained, if existing conditions within many societies do not work against the suggested remedies. Some issues have been left dangling. Is the suggestion to debar genetic information in setting premiums for life insurance a practical step in the direction of justice? Or does it symbolize a willingness to attend to certain causes of inequality while ignoring others? Here pragmatists and idealists differ, the latter fearing that tinkering with the local contexts in which molecular knowledge will be applied is not enough, that deeper social problems will defeat our patched contrivances.

The remaining chapters are concerned with these further questions, questions some will doubtless find abstract and nebulous, while others take them to be the most fundamental and important. I shall start by confronting the bogey of eugenics, which often lurks threateningly behind debates over the future applications of molecular biology. Once the many-sided character of eugenics has been exposed, it will be natural to ask how far we should go in choosing people. Ought we to restrict ourselves to preventing lives likely to be afflicted with disease? How do we delimit disease?

These questions are forced on us by our power, limited at present but probably extensive within a few decades, to decide on the characteristics of future people. More precisely, in the coming years we shall have increasing opportunities to *prevent* certain kinds of lives from continuing. Prenatal testing provides applicable knowledge because prospective parents have the option of terminating a pregnancy. But should that option ever be exercised? We cannot avoid the thorny issue of the moral permissibility of abortion.

Nor can we sidestep questions about the extent to which our genes make us what we are. The history of applications of genetic knowledge reveals how easy it is to find destiny in DNA. If the future is to avoid the oversimplifications and errors of the past, it will be necessary to draw distinctions often disregarded in the enthusiasm for new "scientific" understanding of some trait that fascinates us. We shall need to locate some boundaries in some much-disputed territory.

Other ghosts are more ethereal but no less stubborn. A recurrent nightmare is that increased knowledge of the molecular processes out of which human lives are fashioned will change our conceptions of ourselves. Can we retain the sense of the value of human activities when we discover the minute details of their workings? Will the new biology show that our vaunted freedom is an illusion?

In the two final chapters of this book, I shall attempt to synthesize the individual discussions of these questions and to connect them to the practical problems already considered. The new molecular genetics will make it impossible for us to evade hard choices. Not only shall we have to consider when nascent lives should continue, but there will also be implications for how lives should be supported and how they should end. The medicine of the future will be caught up in consideration of the quality of human lives.

Advances in scientific understanding will require our descendants to work out a moral perspective for applying their new knowledge. Will that perspective have consequences for other social policies? Critics have already challenged what they see as an obsession with molecular medicine that blindly neglects many of the factors that cause human misery. So we shall come back to the debate between pragmatists and idealists. Is it enough to devise piecemeal ways of blocking the definite dangers posed in applying molecular biology? Or are we required, by consistency, to do more, to try to eradicate the underlying social causes of which those dangers are merely symptoms?

All these questions are intricate, and I cannot promise more than ways of thinking about them. I have written this book in the conviction that we cannot neglect either the concrete local problems or the broader more abstract questions. Both perspectives will be needed if the great advances in molecular genetics are to fulfill their promise.

8

Inescapable Eugenics

For over two centuries, first British colonial officials and later the Indian government have struggled to stamp out the practice of female infanticide in rural villages of Northern India. Bound by the caste system, families view daughters as an economic burden and until recently have resorted to crude methods of freeing themselves from such expensive chattels. Baby girls have been killed at birth, usually through asphyxiation or drowning, and even those who are permitted to survive suffer abuse, neglect, and markedly higher mortality rates than their brothers. The arrival of more advanced medical techniques, including amniocentesis, has brought a more humane option. Pregnant women can discover the sex of the fetus and choose to terminate the pregnancy if they find they are carrying a female; in some regions of Northern India, dramatically skewed sex ratios at birth reveal that this has become a popular strategy.

It couldn't happen here, of course—or so you might suppose. Western societies are not governed by prejudices masquerading as religious doctrine, and though progress toward the acceptance of women as equals in the affluent world has been unsteady, imperfect, and incomplete, widespread prenatal testing would probably not issue in *dramatically* unbalanced sex ratios. (Whether it would lead to a preponderance of two-child families, with elder brother and younger sister, is a further question.) But we should not congratulate ourselves too quickly, for we have preferences aplenty. Unlike Caesar, many people do not want to surround themselves with those who are fat. Even people who think it barbaric to persecute homosexuals are often disappointed to discover that a child is gay or lesbian, a fact reflected in the difficulty homosexuals frequently find in telling parents of their sexual orientation. And in many socioeconomic strata in meritocracies, where brains are taken to be key to at least modest levels of security and success, fathers and mothers hope fervently that their children will obtain an average score (or better) on the tests that are supposed to measure intelligence. Genes indicating low IQ or same-sex preference will be very hard to find—and may not exist—but researchers already claim to have found genetic causes of obesity. Although sex selection may be improbable in Western societies, ten years hence prospective parents sharing the attitudes now common may terminate pregnancies because tests disclose the presence of fetal genes indicating obesity, or perhaps genes that are suspected (quite possibly incorrectly) of causing homosexuality or low intelligence.

We have traveled roads like this before. The history of eugenics in Western societies offers a succession of prospects, some whimsical, most dismal, many tragic. Little harm was done by the eugenic exhibits at American state and county fairs, in which proud examples of prize human stock paraded before their neighbors, adorned with ribbons though presumably not decorated with rings through their noses or bells on their toes. Some of the work done at the former Eugenics Record Office at Cold Spring Harbor on Long Island, now transformed into one of the world's major centers in molecular genetics, bears a similar aura of dizziness. Convinced that the large number of naval officers in some families pointed to a hereditary yen for the sea, Charles Davenport, founding director of the Record Office, earnestly sought the allele for what he called "thalassophilia" (literally "love of the sea"), which he took to be a sex-linked recessive expressed only in males.

Most early-twentieth-century eugenic projects had far darker effects. Davenport's office also amassed studies to show the genetic inferiority and undesirability of the peoples from Eastern and Southern Europe. Congress responded by trumpeting the need for racial purity, and quickly transformed rhetoric into action. In 1924 a new Immigration Act greatly restricted the number of people who could enter the United States from the "undesirable" parts of Europe. Combining official immigration policy with the use of intelligence tests, tests whose cultural biases appear dumbfounding in retrospect, Henry Goddard, who pioneered the idea of screening for intelligence at Ellis Island, succeeded in returning to Europe thousands who would be destroyed by totalitarian regimes.

American obsessions with genetic purity were not simply directed at resisting corruption from without. It was also considered important to extirpate the putrescence of homegrown genetic infections spread by the shiftless and degenerate members of the population who often bred in far greater numbers than their respectable (middle-class, white, Protestant) counterparts. In the 1920s eugenicists publicly lamented the relatively low birthrates of graduates from elite universities. Determined to prevent America from being overrun with "the feebleminded," they campaigned for compulsory sterilization laws and won partial victories through legislation to "treat" the inmates of institutions in many states. Even during the 1940s and 1950s, sterilization was sometimes made a condition of discharge from mental hospitals and prisons.

By then, of course, knowledge of Nazi eugenic practices had caused changes in attitude, not only in those countries (such as Britain) in which eugenic pronouncements had not issued in social policy, but also in America. Beginning in 1933, Hitler had introduced compulsory sterilization on a far grander scale than anything enacted elsewhere, using it as the first instrument in the promotion of "race biology." Other tools followed, and by 1939 the Nazis were using more direct means to eliminate those they judged biologically inferior—people diagnosed as suffering from mental disorders, homosexuals, Gypsies, and Jews. Both the brutality of the methods and the patent attempt to portray social prejudice as objective biology appalled the world. "Eugenics" became a term tightly associated with Hitler's profoundly evil practices, a word with so powerful a stigma that it can instantly stop debate.

So when it is suggested that contemporary molecular biology will inevitably contribute to a revival of eugenics, the implications seem clear: We should have none of it. Genetic testing appears benign when it focuses on reducing the incidence of those rare but terrible diseases that afflict children with massive disruptions of development and early death. Some examples have figured in earlier chapters, but there are many others. Children born with Hurler syndrome show decelerated development toward the end of their first year. They typically deteriorate, growing abnormally and losing cognitive functions, and virtually all die before they are ten years old. Children with Sanfilippo syndrome may survive longer, but they are severely retarded, and their aggressive behavior is frequently difficult to manage. Unlike Hurler children, who are usually "placid and lovable," Sanfilippo patients may be wild and unreachable even when in the most tender and informed care. Many surely find the ability to predict these syndromes liberating, enabling prospective parents to prevent the inexorable decay of the Hurler child or the incomprehensible ferocity of children with Sanfilippo syndrome. Yet for at least some readers of the history of eugenics, these are the first steps on the road to a dreadful destination. In this view, once we begin thinking in terms of "innate defects," social pressures will expand the category of genetic deficiencies, and we shall end by cloaking injustice and prejudice in professions of biology, less monstrously than Hitler, or even than Davenport, but nonetheless harmfully enough.

Others harbor different fears. They are moved by fictional portrayals of individuals and societies who meddle with life and who try to shape people according to some distorted vision of the good. They see contemporary molecular biology as moving towards Baron

Frankenstein's laboratory or the Central London Hatchery of *Brave New World*. Today we undertake timid ventures in prenatal testing and gene replacement; tomorrow we shall dispassionately design children and "decant" them into a ruthlessly planned world that has lost its humanity.

Because fine-grained genetic engineering is still remote and may be impossible, fear of Frankenstein is easier to dismiss than anxieties about repeating the errors of our eugenic past. At present, indeed for the foreseeable future, we cannot select *for* the human traits we deem desirable, shaping people according to our ideals, but we can select *against*, terminating those pregnancies in which fetuses bear unwanted alleles. Northern Indians already do it in a very particular way. Members of affluent societies will soon have the opportunity on a large scale. No doubt, initially, they will be moved by concern for the misery associated with the most devastating diseases—but where will they stop?

Eugenics was officially born in the writings of Charles Darwin's cousin, Francis Galton, who campaigned for applying knowledge of heredity to shape the characteristics of future generations. In retrospect, we can recognize the new theoretical science as a mixture of a study of heredity and some doctrines about the value of human lives. Galton's approach to studying inheritance, by looking for statistical features of the transmission of phenotypes, was original and was discarded by the growing number of his eugenic descendants who embraced Mendelian genetics. With characteristic Victorian confidence, however, Galton did not offer a critical discussion of the values underlying his judgments about proper and defective births. Assuming that his readers would agree about the characteristics that should be promoted, he set about the business of promoting them.

Separated from Galton by over a century, we can see how eugenic judgments have mixed science with the values of dominant groups and also how the prejudices have been so powerful as to distort scientific conclusions: Men with a mania for eradicating "feeblemindedness" convinced themselves that there must be genes to be found and duly "found" them. A fundamental objection to eugenics challenges the presupposition that there is *any* system of values that can properly be brought to bear on decisions about genetic worth. Galton's Olympian confidence that he could decide which lives would best be avoided easily provokes the reaction that we should abandon the pretense of being able to judge for others. People with severe disabilities who have attended workshops for human geneticists sometimes pose the question forcefully in words, sometimes even more vividly in their presence and determination: Who are you to decide if I should live?

Yet even as we admire those who have overcome extraordinary adversity to make rewarding lives for themselves, we should remember the dreadful clarity of some examples of genetic disease. Those who watch the inevitable decay of children born with Hurler syndrome or Canavan's disease, those who see the anguish of parents as they care for children whose genes prevent development beyond the abilities of an infant have no difficulty in deciding that similar sufferings should be avoided. In the spring of 1994, at a public discussion of the impact of the Human Genome Project in Washington, D.C., a man in late middle age protested the tendency to see only the problems of the Project, relating how his daughter had given birth to two children with neurofibromatosis. His tone, not his words, conveyed the grief of his family as well as his conviction that abstract fears of eugenic consequences should not block attempts to spare others similar agonies.

As a theoretical discipline, eugenics responds to our convictions that it is irresponsible not to do what can be done to prevent deep human suffering, yet it must face the challenge of showing that its claims about the values of lives are not the arrogant judgments of an elite group. Of course, if eugenics were *simply* a theoretical discipline, pursued by Galton's successors in their studies, there would be little fuss. Precisely because some are concerned about a revival of eugenic *practices*, while others fear that the label "eugenics" will be misapplied to humane and responsible attempts to eliminate pain and grief, questions about the eugenic implications of contemporary molecular biology have more than academic interest. Unless we look past the swastika and achieve a clearer picture of eugenics in action, these important questions will prove irresolvable.

Exactly when are people practicing eugenics? The Nazi doctors, the Americans worried about "racial degeneracy," would-be social reformers like Sidney Webb and George Bernard Shaw, and peasant families in Northern India, do not agree on very much, but all of them hope to modify the frequency with which various characteristics are present in future populations. Eugenic practice begins with an intention to affect the kinds of people who will be born.

Translating that intention into social action requires four types of important decision. First, eugenic engineers must select a group of people whose reproductive activities are to make the difference to future generations. Next, they have to determine whether these people will make their own reproductive decisions or whether they will be compelled to follow some centrally imposed policy. Third, they need to pick out certain characteristics whose frequency is to be increased or diminished. Finally, they must draw on some body of scientific information that is to be used in achieving their ends. Practical eugenics is not a single thing. Human history already shows a variety of social actions involving four quite separate components, each of which demands separate evaluation.

Introducing contemporary molecular biology into prenatal testing will lead us to engage in *some* form of eugenics, but that consequence, by itself, does not settle very much. For it is overwhelmingly obvious that some varieties are far worse than others. Greater evils seem to be introduced if we move in particular directions with respect to the four components: More discrimination in the first, more coercion in the second, focusing on traits bound up with social prejudices in the third, using inaccurate scientific information in the fourth. Unsurprisingly, Nazi eugenics was just as bad as we can imagine with respect to each component. The Nazis discriminated among particular populations for their reproductive efforts, selecting the "purest Aryans" for positive programs, using "special treatment" on groups of "undesirables." Starting with compulsory sterilization, they proceeded to the ultimate form of coercion in the gas chambers. The repeated comparison between Jews and vermin and the absurd—but monstrous—warnings about the threats to Nordic "racial health" display the extent to which prejudice pervaded their division of human characteristics. Minor, by comparison, is the fact that much of their genetics was mistaken.

Scientific inaccuracies infect other past eugenic practices in ways that appear more crucial. Henry Goddard's efforts to keep America pure led him to administer intelligence tests to newly arriving immigrants; even the staunchest contemporary advocate of IQ would have difficulty defending Goddard's assumption that his tests measured "innate hereditary tendencies." Those cast up at the foot of the Statue of Liberty found that their inability to produce facts about the recent history of baseball indicated their lack of native wit. Likewise, decisions to sterilize the inmates of state institutions often rested on abysmally poor evidence. Carrie Buck and her sister Doris were victims of the zeal to dam up the feebleminded flood, and it was Carrie's case that provoked one of the most chilling lines in the history of Supreme Court decisions. In 1927, finding in favor of the lower court decision in *Buck* v. *Bell*, Justice Oliver Wendell Holmes pronounced that "three generations of imbeciles are enough."

The three generations of imbeciles were three members of the Buck family. Carrie Buck's mother had been diagnosed as "feebleminded," Carrie herself had been placed in the Virginia Colony for Epileptics and Feebleminded; on the basis of a Stanford-Binet intelligence test, she was assigned a mental age of nine. The third generation consisted of her illegitimate daughter, Vivian, seven months old at the time that the original decision for sterilization was made. Because a Red Cross worker thought that she had "a look" about her and a member of the Eugenics Record Office claimed, on the basis of a test for infants, that she had below-average intelligence, Vivian too was classified as feebleminded. Three generations of feeble-mindedness demonstrated that the defect was hereditary.

At least the classifications of Carrie Buck and Vivian were quite erroneous. Vivian died while still a young child, but she had completed second grade and had impressed her teachers. In 1980, when she was in her seventies, Carrie Buck was rediscovered and was visited by doctors and scholars concerned with the history of sterilization laws. They found an ordinary woman who read the newspaper daily and who tackled crossword puzzles. The central figure in the tragic story was no "imbecile," not even according to the technical criterion which eugenic enthusiasts employed to grade the feeble-minded (imbeciles were adults with a mental age between six and nine).

Failure to distinguish the components of eugenic practice blurs our vision of the injustices that have been done. What is the real moral of the case of Carrie Buck? Not simply that the judgment was mistaken, that Carrie and her sister Doris, both of whom were sterilized, did not carry genes for "feeblemindedness." Besides the scientific error, the practice of compulsory sterilization also destroyed something fundamental to people's lives. Even if Carrie, Doris, and Vivian had borne genes that set limits to their mental development, should they have been forced to give up all hopes of bearing children? Like the Nazis, albeit on a far smaller scale, American eugenicists carried out a coercive practice: They compelled some individuals to follow a social policy that was divorced from any aspirations that those who were treated may have had. Doris lived outside the asylum, married, and tried to have children. Only much later did she understand what had been done to her.

The brutal compulsion of the Nazi eugenics program prompted an important change in postwar efforts to apply genetic knowledge. Everyone is now to be her (or his) own eugenicist, taking advantage of the available genetic tests to make the reproductive decisions she (he) thinks correct. If genetic counseling, practiced either on the limited scale of recent decades or in the much more wide-ranging fashion that we can anticipate in the decades to come, is a form of eugenics, then it is surely *laissez-faire* eugenics. In principle, if not in actuality, prenatal testing is equally available to all members of the societies that invest resources in genetic counseling. Ideally, citizens are not coerced but make up their own minds, evaluating objective scientific information in light of their own values and goals. Moreover, the extensive successes of molecular genetics inspire confidence that our information about the facts of heredity is far more accurate than that applied by the early eugenicists. As for the traits that people attempt to promote or avoid, that is surely their own business, and within the limits of available knowledge, individuals may do as they see fit. Laissez-faire eugenics, the "eugenics" already in place and likely to become ever more prominent in years to come, is a very different form of eugenics from the endeavors of Davenport, Goddard, and Hitler's medical minions.

Identifying the gulf between laissez-faire eugenics and the horrors that underlie the stereotype makes room for discussing the important questions surrounding applications of molecular genetics but does not resolve them. Banning prenatal tests by tagging them with the ugly name "eugenics" should not substitute for careful thought about their proper scope and limits. Everything depends on the *kind* of eugenics we practice.

We know that some genetic conditions cause their bearers to lead painful or truncated lives in all the environments that we know how to arrange for them. We know also how to identify, before a fetus is sentient, whether or not the fetus carries one of those conditions. Naively, we might try to avoid the smear of eugenics by insisting that nobody should use this information for selective abortions—we shall not interfere with the genetic composition of future populations. But once we have the option of intervening, this allegedly "noneugenic" decision shares important features with eugenic practices. Tacitly, it makes a value judgment to the effect that *unplanned* populations are preferable to *planned* populations. More overtly, it imposes a bar on decisions that individuals might have wished to make, depriving them of the chance to avoid great future suffering by terminating pregnancies in which fetuses are found to carry genes for Sanfilippo syndrome, neurofibromatosis, or any of a host of similarly devastating disorders. When we know how to shape future generations, the character of our descendants will reflect our decisions and the values that those decisions embody. For even if we compel one another to do nothing, that is to judge it preferable not to intervene in the procreation of human life, even to subordinate individual freedom to the goal of "letting what will be, be."

Molecular knowledge pitches us into some form of eugenic practice, and laissez-faire eugenics looks initially like an acceptable species. Yet its character deserves a closer look.

The most attractive feature of laissez-faire eugenics is its attempt to honor individual reproductive freedom. Does it succeed? Are the resources of prenatal testing in affluent societies equally open to all members of the population? Do they help people to make reproductive decisions that are genuinely their own? And is that really a proper goal? Since individual reproductive decisions have aggregate consequences for the composition of the population, should there not be restrictions to avoid potentially disastrous effects? Finally, because individual decisions may be morally misguided—as with those who would select on the basis of sex—will laissez-faire eugenics foster evil on a grand scale?

These serious questions emerge once we have appreciated the dimensions along which eugenic practices may be evaluated, for they correspond to three of the four components: discrimination, coercion, and division of traits. (Concern with the status of our molecular knowledge is less urgent when we concentrate on physiological characteristics, but will reappear later when we consider beliefs about the genetic basis of behavior.) Discrimination and coercion are prominent features of the history of eugenic excesses; only the gullible should believe that simple declarations that genetic testing will be open to all and free from social directives are enough to ensure that history will not repeat itself in these respects. Most obviously in the United States, but possibly in countries that already assure their citizens access to medical care, the costs of such medical technologies as prenatal tests and *in vitro* fertilization may prove an effective barrier to their broad availability: The poor may lack options that the affluent exercise. Predicting the likely consequences of these kinds of inequalities is not hard. If prenatal testing for genetic diseases is often used by members of more privileged strata of society and far more rarely by the underprivileged, then the genetic conditions the affluent are concerned to avoid will be far more common among the poor—they will become "lower-class" diseases, other people's problems. Interest in finding methods of treatment or for providing supportive environments for those born with the diseases may well wane.

The fault lines that run through our societies may threaten future prenatal decisions in other ways as well. Laissez-faire eugenics promises to enhance reproductive freedom. Prenatal testing is to provide unprecedented choices for prospective parents—and, indeed, it will sometimes spare mothers and fathers the anguish of watching young children degenerate and die. Yet the proclamation of reproductive freedom does not necessarily translate into increased autonomy in the clinic. Although the storm trooper's gun is the least subtle of a variety of ways in which social values may shape individual decisions, simply avoiding official decrees of "racial health" and announcing that prospective parents' decisions are to be their own does not ensure that people will act in ways that correspond to their most fundamental ideals.

There have already been women and couples for whom the bad genetic news has been doubly agonizing. Discovering that a fetus has an extra copy of chromosome 21—that it has Down syndrome—they know that the future person would have a more limited life.

However, they also know that, with love, nurture, and support, people with Down syndrome can sometimes defy the gloomy predictions that used to lead some doctors to treat the extra chromosome as a marker for abortion. They are prepared to provide the love and nurture, but they cannot count on the support. For some, it is a matter of economics, of not being able to pay for special programs; for others, the problem stems from the attitudes prevalent in the segment of society they occupy, a tendency to view "mongol" children as defective and to write them off at birth. So, moved by concern for the future happiness of the child who would be born, they decide—reluctantly and against their own deep commitments—to terminate the pregnancy.

Individual choices are not made in a social vacuum, and unless changes in social attitudes keep pace with the proliferation of genetic tests, we can anticipate that many future prospective parents, acting to avoid misery for potential children, will have to bow to social attitudes they reject and resent. They will have to choose abortion even though they believe that a more caring or less prejudiced society might have enabled the child who would have been born to lead a happy and fulfilling life. Laissez-faire eugenics is in danger of retaining the most disturbing aspect of its historical predecessors— the tendency to try to transform the population in a particular direction, not to avoid suffering but to reflect a set of social values. In the actual world unequal wealth is likely to result in unequal access, and social attitudes will probably prove at least partially coercive. How are these problems to be solved?

Not by jettisoning the use of prenatal testing entirely. Parents who have seen their own children grieve over a child with neurofibromatosis or Hurler syndrome, who have watched marriages torn apart, bright lives quenched, rightly remind opponents of prenatal testing of the tragedies that the tests may prevent. Only the callous would refuse to allow some to benefit from the new resources of molecular genetics simply because those resources are not available to all. But even though unequal distribution might be tolerated temporarily, societies that introduce prenatal testing have a moral obligation to work toward making it available to all their citizens.

Similarly, if laissez-faire eugenics is to accord with the advertisement that it promotes individual reproductive freedom, our societies will have to combine attempts to bring into the world children whose lives are not sadly restricted with public commitment to assist those who are born to realize the highest possible degree of development. Disability activists already fear that the spread of prenatal tests will erode the tenuous systems of support that have made it possible for many people to go far beyond the limits once foreseen for them. If those systems decay, prospective parents will experience an ever more relentless pressure to eliminate those whom their society views as "defective." To make their decisions maximally free, they need to know not only how the genetic condition of the fetus would affect the life of the person who would be born, given the range of manageable environments, but also that their society is committed to helping them bring about an environment in which that life will flourish insofar as it can. Only if they are assured that all people have a serious chance of receiving respect and the support they need can prospective parents decide on the basis of their own values.

Social change might make it possible for laissez-faire eugenics to live up to its billing, fastidiously promoting the reproductive freedom of all. But is that really desirable? Prenatal decisions do not affect only the parents; they have consequences, very directly, for the cluster of cells within the uterus and, more remotely, for other members of society who may have to contribute to support for the child who is born.

Those who believe that the cluster of cells already counts as a person view prenatal testing as a means of inspiring evil: Laissez-faire eugenics would hand out licenses to murder. Others do not oppose the taking of fetal lives in principle but doubt that prospective parents will make socially responsible decisions. One by one, the effects of terminating or continuing a pregnancy may be small; in the aggregate, they can make profound differences to the lives of our descendants.

If people proceed myopically, guided by their own dim lights, the consequences may prove disastrous. Seventy years ago unregulated breeding was widely assumed to lead to genetic catastrophe. Eugenicists everywhere suggested that *Homo sapiens* would be buried under a "load of mutations" and, more chauvinistically, Anglo-Saxons lamented "the passing of the great race." Today the concerns are expressed less stridently, more apologetically, and in terms of the *economic* impact on societies. Perhaps the commitment to fostering the full development of those born with genetic disabilities—a commitment which seems morally unassailable—would combine with shortsighted individual decisions to produce impossible obligations. Prospective parents, confident that their society will support a child with a genetic disorder, solemnly take on the responsibilities of providing love and care. As the product of their thoughtful, moral decisions, large social resources—resources that might have been used to promote the welfare of many other children—are consumed in compensating for the genetic misfortunes of their offspring.

In the presence of practical constraints, attractive ideals conflict. Reproductive freedom is important to us; providing support for all members of society, including people with expensive genetically imposed needs, is equally important. But our resources are finite. In a callous society, as we have seen, individual reproductive freedom is severely constrained. In a caring society individual reproductive freedom may lead to social disaster.

Not all chapters in the history of eugenics are unremittingly bleak, and the reflections of some earlier thinkers indicate a potential solution to our dilemma. British social reformers—George Bernard Shaw, Sidney and Beatrice Webb, and others who followed them—hardly correspond to the stereotypes of eugenic repression: They believed that eugenics, practiced freely, would be part of a systematic scheme of social reform. The society they envisaged would provide support for the full development of all its members and would use eugenic methods to ensure a population in which this commitment could be met. But they did not intend to dictate reproductive policy. How, then, were the reformers to guarantee that *right* choices would be made, that parents would procreate to achieve, collectively, a population of progeny who could be fully supported by the available resources? Their answer emphasized eugenic education. People would make the right decisions because they would understand the consequences of their decisions, both for their offspring and for society.

Today's enthusiasts for the use of molecular genetics in prenatal testing rarely call themselves "eugenicists," but their vision of future reproductive practice depicts a particular version of laissez-faire eugenics—*utopian eugenics*—which is remarkably similar to the ideas of Shaw and his friends. Utopian eugenics would use reliable genetic information in prenatal tests that would be available equally to all citizens. Although there would be widespread public discussion of values and of the social consequences of individual decisions, there would be no societally imposed restrictions on reproductive choices— citizens would be educated but not coerced. Finally, there would be universally shared respect for difference coupled with a public commitment to realizing the potential of all those who are born.

Who's afraid of utopian eugenics? Some critics surely fear that utopian eugenics deserves its name, that the conditions it requires cannot be sustained in any society. They might well concede that, in some attenuated sense, benign applications of molecular genetics are possible while maintaining that the results in practice will be far more disturbing. Has any society succeeded in giving its citizens reasonable access to medical resources, or in providing the basic support that all its children need to realize their potential? Can education be expected to succeed in promoting responsible reproductive decisions? Is it possible that an educational program would enhance reproductive freedom, rather than collapsing into a system of ideology that would reinforce widely held prejudices? If we are indeed committed to some form of eugenics, then utopian eugenics seems the most attractive option. But these questions will need to be answered if we are to forestall the worry that it will inevitably decay into something darker.

The chapters that follow address even more fundamental concerns about utopian eugenics. If prenatal tests are to be employed as a prelude to ending nascent human lives, then abortion must, at least sometimes, be morally defensible. Moreover, our descendants will have to arrive at a clearer conception of the ways in which genotypes influence human phenotypic traits: Repeating the determinist errors of earlier versions of eugenics would be truly tragic. We should also reflect on how the enterprise of choosing people changes our conceptions of human life and human freedom and of their value. If I am right in my diagnosis, human molecular genetics is already committed to that enterprise, and before we are carried onward by its momentum, there are important steps to take.

Most basic is the task of clarifying utopian eugenics, uncovering the considerations that should guide our reproductive choices. To envisage a practice in which reproductive decisions are both free and educated returns us to the problem with which this chapter began. The trouble is to identify the content of the education, to say what kinds of fetal characteristics would properly lead responsible people to terminate a pregnancy. In thinking about the proper form of eugenics, we have focused thus far on three important components: accurate information, open access, and freedom of choice. The issue of trait discrimination has figured only as an afterthought to the insistence on reproductive freedom: Laissez-faire eugenics allows people to make up their own minds about which traits to promote, which to avoid. Yet matters have proved to be not quite so simple. Utopian eugenics proposes that there should be some encouragement to draw the distinction in a particular way. Which way? The Northern Indians, we believe, do not have it right, whereas the doctors who helped reduce the frequency of Tay-Sachs recognize part of the truth. But exactly where between these polar cases is the line to be drawn?

There is an obvious answer. Abortion is appropriate when the fetus suffers from a genetic disease. Preventing disease has nothing to do with imposing social values, for whether or not something is a disease is a matter of objective fact.

Ultimately, this answer will prove inadequate, but understanding its difficulties will point us toward something better. Once we have left the garden of genetic innocence, some form of eugenics is inescapable, and our first task must be to discover where among the available options we can find the safest home.

9

Delimiting Disease

The Modern Civics class, taught by a young woman of serene
warmth, uses the most up-to-date materials, including the fifteenth
edition of the famous text *Your Reproductive Responsibilities,* published just last year in 2069. The teacher starts with the darkness of
prehistory, the days before prenatal testing was available to responsible citizens. Once, she tells her students, many babies were doomed
to die in infancy, there were special institutions for "defective" children, and the more enlightened nations diverted large sums from
other health and education projects to provide special care for children with genetic disabilities. But the progress of the reproductive
responsibility movement has been heroic: Tay-Sachs is a thing of the
past, Down syndrome is virtually eliminated, congenital forms of
heart disease are now extemely rare, there are far fewer people with
mutant tumor-suppressor genes, far fewer fat people, far fewer homosexuals, far fewer short people. All this is the work of molecular
geneticists, doctors, counselors, and—she smiles modestly—of the
teachers who have helped teenagers recognize their reproductive responsibilities.

A student raises her hand; she has heard that "mental defectives"
are still sometimes born, that not everybody takes the tests or acts
appropriately. Indeed, the teacher sadly acknowledges, education

sometimes fails to convey the important message, and sometimes people who have sat through the class form different opinions; we must respect these differences and not condemn those who do not behave as we would have done. But the student persists; she has heard stories of people who went too far, who terminated pregnancies because the fetus was female, even, in one case, because it carried a gene for dark eyes. Unfortunate but true, the teacher admits, using the question as an opportunity to present the historical data on uses of amniocentesis to select for sex. The class agrees that this is barbaric; their society would never countenance any such practice.

With student interest high, the teacher raises a question of current controversy. The fifteenth edition of the text already embodies the accepted ideas that homosexuality, mutant tumor-suppressor genes, and obesity have personal and social costs, but debate rages over what the planned sixteenth edition should say about left-handedness, whose genetic basis is now thoroughly understood. Left-handers can function quite normally—so long, of course, as special support is available—but they continue to have reduced life expectancy. One student points out that the resources used to make the world safe and easy for left-handers could be channeled toward different projects, improvements in support for higher education, for example. The students are moved by his argument, and though the class is divided, a majority believes that responsible procreators would promote dexterity and avoid the sinister.

Was medicine meant for this? When amniocentesis was introduced into Northern India, no doubt it was hoped that the incidence of the more common genetic diseases and disabilities would decrease. Being female is common enough, genetic enough (at least in one sense of "female"), but neither a disease nor a disability. Using prenatal testing to select for sex violates the Hippocratic conception of medicine, twisting biomedical technology to serve social prejudice instead of promoting health and life. If our descendants employ the genetic tests spawned by the new molecular biology to try to produce only straight, skinny, right-handed progeny, then, whether or not they are successful, they too will have turned their backs on Hippocrates.

Utopian eugenics needs a boundary to circumscribe those kinds of prenatal tests that the morally sensitive, well-informed citizen will use—and to demarcate them from the practices of unenlightened eugenicists. A natural proposal, one that surfaced earlier in considerations of gene replacement therapy, is to single out those conditions that constitute diseases as the appropriate targets for testing and termination. The official aims of Western medicine—to promote health and to cure disease—appear comprehensible: We can easily convince ourselves that we know what diseases are, and armed with that concept, we can readily define health as the absence of disease. Yet few people, few doctors even, encountering some latter-day Socrates, could meet the challenge of providing a satisfactory definition. Perhaps that does not matter, for we surely know one when we see one, we can tell easily that cancer, multiple sclerosis, pneumonia, and cystic fibrosis are diseases, and that being female or left-handed is not, so we can draw a line between interventions to ameliorate the former conditions, proper parts of medicine, and sex selection based on prenatal testing, an abuse of medical technology. The challenge to define "disease" can be ignored.

Or can it? From 1900 to 1960 many homosexuals believed that it was better to be counted as sick than as criminal, and they welcomed the official (American) psychiatric classification of homosexuality as a disease. During the early 1960s, after a bitter debate, homosexuality was removed from the list of diseases in the American Psychiatric Association's *Diagnostic and Statistical Manual.* Psychiatry corrected a well-intentioned mistake born originally of the recognition that a group of social pariahs were not, after all, children of Sodom or Gomorrah but still infected with the idea that their undesirable ("disgusting") activities must testify to immorality or sickness. Thirty years later we can easily understand how widespread social prejudices narrowed the range of apparent options. Are we immune to kindred errors? When we confidently pronounce that some conditions are diseases and others are not, are we demonstrating our ability to know them when we see them, or are we merely broadcasting social values that are so widely shared that they are never called into question?

Telling hawks from handsaws is easy enough, and the vast differences between leukemia and myotonic dystrophy, on the one hand, and being female and having brown eyes, on the other, refutes the extreme doctrine that *all* disease classifications express our social values. Nonetheless, there is ample room for debate about the status of a variety of conditions, some behavioral, others physiological. Systems of education throughout the affluent world enshrine cognitive gradations: There are institutions for those who are "severely retarded," special schools for children with "developmental disabilities," special classrooms for the "learning disabled," special programs to help those who learn at a slower rate than their classmates. Surely not all these children are diseased, not all are appropriate targets for medical interventions, but the lines of demarcation are very hard to draw. Human height and body build display a similar range of variation. At the extreme are people with hormonal disorders, who balloon into obesity no matter how firmly they follow prescribed diets. In the campaign for achieving a socially approved shape, there are many millions who lose the battle with their own metabolism. Do those who "can't look at food without gaining weight" suffer from disease? Is the fashion for diets approved by doctors a sign that medicine is fulfilling its proper responsibilities?

Answers to these questions take one of two general forms. Some scholars, *objectivists about disease,* think that there are facts about the human body on which the notion of disease is founded, and that those with a clear grasp of these facts would have no difficulty drawing lines, even in the challenging cases. Their opponents, *constructivists about disease,* maintain that this is an illusion, that the disputed cases reveal how the values of different social groups conflict, rather than exposing any ignorance of facts, and that agreement is sometimes even produced because of universal acceptance of a system of values—as, for example, psychiatrists once agreed that homosexuals were sick or, today, we might count some obese people as suffering from hormonal disorders. The fear made vivid in the opening fantasy—Modern Civics 2070—far-fetched though it may appear, is that no line can be drawn, that objectivism about disease cannot be sustained, that decisions about prenatal testing must ultimately turn on a social consensus about what kinds of lives are valuable. If constructivists are right, if disease classifications reflect the ways in which societies express their idiosyncratic preferences, then a "utopian" eugenics that boasted of its commitment to preventing disease would be of a piece with the darker chapters of the enterprise of choosing people. At bottom, there would be no more objective basis for terminating pregnancies when the fetus is found to carry the Sanfilippo genotype than for striving to eradicate lesbianism or obesity. Constructivism challenges us to isolate a value-free, objective notion of disease, or else to find some other way of circumscribing the proper scope of prenatal interventions.

Objectivists recognize that striking and central examples of disease share common features. Disease often involves pain, sometimes brings a limitation of activities, sometimes results in premature death, and is not statistically normal. If we reflect on a wider variety of diseases, as well as on conditions that are not diseases, it quickly becomes apparent that, singly or in combination, the characteristics of the list will not serve to demarcate disease. There are diseases without pain—hypertension deserves at least part of its billing as the "silent killer"—as well as painful conditions, like teething, labor in childbirth, and the sensations athletes strive for in training, which are not diseases. Social and natural processes, such as incarceration and aging, limit our activities more than do many diseases, while on a worldwide scale war and famine contribute enormously to premature death. Since few people have red hair, or live past one hundred, there are uncommon conditions that are not diseases, and conversely, tooth decay used to be almost universal. Objectivists must search out more subtle factors that underlie the variety of things we call diseases.

Contemporary medicine, partitioned into specialities and subspecialities, provides a useful clue for developing more sophisticated objective criteria. Experts typically focus on normal functioning and on breakdowns in function of particular organs and systems of the body: There are specialists in cardiology, opthalmology, renal disorders, and so forth. Human bodies are healthy when each of their constituent subsystems does the job it is supposed to do. Diseases result from malfunctioning of one or more of the parts. Returning to the spectrum of examples that caused trouble for simpler forms of objectivism, it is not hard to align the diseases with instances of functional breakdown: Tooth decay and hypertension embody losses of function in teeth and arteries, respectively; redness does not interfere with the function of hair; if anything, the pains felt by athletes are functional; and so it goes.

With this approach objectivists shift the burden of defining disease to one of understanding the notion of normal functioning (or, equivalently, of breakdown in function). Metaphors come easily: The function of an organ is the job it is supposed or intended to do, the task for which it was designed. To sustain these metaphors without introducing social values is another matter. If normal functioning is simply making it possible for people to act in ways that accord with social standards of valuable activity, then the route through functions will lead back only to constructivism. Is there an objective, value-free notion of function that can be used to delimit disease?

Biology is full of references to function. Ever since William Harvey, we have known that the function of the heart is to pump the blood. Seventeenth-century doctors and physiologists thought they could understand functions literally in terms of design: God intended human hearts to pump blood, and God's intention bestows this function.

Whether or not they are numbered among the faithful, few if any contemporary biologists would subscribe to this theological conception of function. For some, the design of the Creator has given way to the action of natural selection; for others, functions are identified in terms of contributions to the most general goals of organisms.

On many mornings, as I run my puffing way along the edges of our local park, I startle some of the local jackrabbits, inspiring them to bound off into the chapparal. Their long ears contrast with those of the rabbits of my soggy English youth, but the jackrabbits of semiarid Southern California are well served by those appendages: Their function, of course, is to dissipate heat, a job which their European cousins rarely need to perform. On one account, this ascription of function simply records the way in which natural selection shaped the jackrabbit's ears. Ancestral rabbits with longer ears were better able to cool themselves, maintain a constant body temperature, remain active longer, forage more effectively, and so thrive in the evolutionary business of survival and reproduction. On the rival story, the goal of survival is simply given, and the rabbit's body is viewed as a complicated machine needing to be maintained at roughly constant temperature if that goal is to be achieved. Long ears enable excess heat to be lost, making it easier for rabbits in hot, dry climates to keep their body temperatures within an acceptable range and so contributing to their survival.

Darwin's world is one in which organisms struggle against hostile forces and functions are identified by noting the particular ways in which they are successful. Grounding functions in adaptations or in contributions to survival provides an objective notion of function, generating a value-free notion of breakdown of function and, consequently, an objectivist conception of disease.

Returning, finally, to the problem dramatized by Modern Civics 2070, we can give a principled restriction of biomedical technology that will distinguish proper medical uses from abuses born of social prejudice. Prenatal testing and selective termination of pregnancies should be confined to those cases in which the presence of certain alleles causes, in the known environments, some breakdown of normal functioning.

However, once the bits and pieces of this multilayered story have been assembled, the popular response—"If we stick to curing disease and promoting health, all will be well"—begins to lose its attraction. Parts of our bodies are very probably adapted to promote our performance in the evolutionary competition in all kinds of ways that matter little to us. More to the point, some of our principal concerns about our bodies may be quite remote from, even at odds with, inherited human tactics for increasing our reproductive success or our chances of survival. One leading hypothesis about the increase in breast cancer in affluent societies proposes that women's risks vary directly with the length of time between menarche and first pregnancy. If human breasts are adapted for feeding infants relatively early in life, delaying childbirth is profoundly dysfunctional, affecting both a woman's reproductive output and her chances of survival. But many Western women do not want to restore "normal" functioning, hoping instead that medicine will find ways to avoid the increased risks of death without employing the tactics with which human evolutionary history has equipped them.

From the Darwinian perspective, functional traits are those that promote reproductive success, that bestow on their bearers the ability to bequeath more copies of their genes to future generations than they would otherwise have managed. Compulsive sperm donors aside, most people are not overwhelmingly concerned with the spread of their genes. Few would sacrifice crowded hours of glorious life for the consolation of knowing that they would have greater genetic representation in future generations. Of course, the normal functioning of our bodies typically does contribute to our well-being and to pursuit of the things we value. Attention to other organisms plainly reveals, however, that the promotion of survival and reproduction may come apart. In some species—for example, salmon and many kinds of insects—maximizing reproduction involves dying: "Successful" organisms explode in a blaze of procreative triumph. Members of the species with nonfunctional reproductive systems might actually live longer.

Speculative examples can be found closer to home. Imagine that the recent suggestion that some male homosexuals have smaller hypothalamic nuclei than heterosexual males is confirmed by further research and, indeed, that we discover that nuclei of a particular size range always cause the men who have them to be exclusively homosexual. (This would be quite compatible with the idea that some, many, or most male homosexuals have normal-sized nuclei; the claim that all men with the small nuclei are homosexual should not be confused with the quite different thesis that all male homosexuals have small nuclei—any more than the assertion that all popes are Catholic should be confounded with the doctrine that all Catholics are popes.) Nuclei of the smaller size appear to be dysfunctional, in the obvious sense that they would seem to cause their bearers to have fewer progeny. Yet even if this were so, would it be legitimate to intervene medically, to try to restore "normal functioning"? Are untreated men with the small nuclei "diseased"?

Even though we have been shaped by natural selection, we have non-Darwinian values, so that longevity and fecundity do not assume overriding significance for us. Human evolutionary history has equipped us both with the capacity for culture and with the ability to reflect on our predicaments and choices, and that combination loosens the connection between human valuation and natural selection, between what we want and what is good for our survival and/or reproduction. Once this point is appreciated, we can see why the popular idea of using genetic interventions to restore normal functioning, only normal functioning, and nothing but normal functioning, is mistaken. In attempting to combat the constructivist conception of disease, it is natural to turn to the notion of function and to try to understand functions as tactics used against Darwin's hostile forces. However, laborious attempts to develop an objective—value-free—

notion of disease are ultimately of no help, because they ground our understanding of health and disease in facts about our evolutionary history that, although they may sometimes be congruent with our goals for ourselves, are quite external to those goals.

Our original problem was to understand the proper sphere within which the new molecular technologies should operate, to find where medicine ends and social prejudice takes over. Prenatal testing for Hurler syndrome or neurofibromatosis appears humane and liberating; using amniocentesis to select for sex or curly hair seems morally dubious, to say the least. Whatever its uses, a value-free notion of disease does not give an order of priority for the situations in which it is appropriate to restore normal functioning. There are surely some breakdowns in functioning that cry out more loudly for treatment than others: It is more urgent to respond to myotonic dystrophy or cystic fibrosis than to albinism or to an allergy to hazelnuts. Some conception of value is required to grade the relative urgency of demands for intervention. In light of our reflections on the gap between the Darwinian values that lie behind the notion of normal functioning and people's everyday aims for themselves, we ought to ask why the criterion for urgency should be restricted to cases that involve loss of normal functioning and why all these cases have priority over instances that might be highly ranked by the urgency criterion but which do not involve loss of normal functioning. Instead of starting with all instances of loss of normal functioning and proceeding in order of decreasing urgency, why should we not simply respond to the more urgent needs first, whether or not loss of functioning is involved?

The danger, of course, is that confronting issues of value directly will plunge us into insoluble controversies about where legitimate uses of molecular technology end. Many discussions of the limits of gene replacement therapy or prenatal testing seem haunted by the fear that, if the relevance of judgments about values is once admitted, then there will be no basis for condemning whatever practices of genetic interference people choose to pursue. Attempts to decrease the incidence of Tay-Sachs will be no different from Northern Indian sex selection or from anticipated future efforts to "cure" lesbians. Yet is the fear well founded? Is it true that if value-free demarcations are dead, then everything is permitted?

Few people have trouble with some decisions about medical priorities. Faced with the options of investing equal resources either in preventing the mental retardation associated with untreated PKU or in curing those with allergies to hazelnuts, nobody experiences pangs of difficult deliberation. Even the constitutionally unstoical (like me) can endure mild allergies without serious diminution in the quality of their lives. By contrast, the child who suffers mental retardation because of the buildup of unmetabolized phenylalanine (and the related low levels of tyrosine) has a life whose range of opportunities has been dramatically narrowed. There are many things she might otherwise have been—a teacher, a mother, a lover, an office worker, a close friend, an athlete, a factory worker, a political activist—but she will be none of these. She would surely have experienced pain as well as pleasure, frustration as well as exhilaration, and it is even possible that the overall balance of pleasurable experiences in her life would have been less than those she receives within the sheltered institution in which her days drift by. Nevertheless, because she has never had the chance to make those choices that would have defined her as a person, her life has been sadly diminished—indeed, it is tempting to say that she has never lived at all.

Often it is a mark of tolerance to recognize that others think differently and that their values diverge from our own. Tolerance is usually appropriate, but not always. A scheme of values that measured the loss incurred by inability to eat hazelnuts as greater than that resulting from severe mental retardation would not simply be different; it would be obscene. Human lives vary in their quality, some are full of pain or pathetically restricted in the choices they allow, others are happier and richer. To overlook the moral significance of these variations, to be bewitched by the alleged relativity of values, to be swept into the constructivist account that sees no difference between eradicating Tay-Sachs in North America and extirpating the females of Northern India ("these are our values, those are theirs"), is to lose sight of our capacity at least to make *rough* judgments—rough judgments that are not merely the expression of local tastes—about the urgency of medical interventions. By understanding the bases of these judgments, we may hope to refine our ability to assess more complicated cases and eventually to identify the limits of the proper uses of biomedical technology.

The impetus to intervene should come from recognizing the ways in which the quality of a life can be adversely affected. Objectivists about disease think of medicine as focused on the functions of parts of the body and as attempting to restore normal functioning. Constructivists about disease think of all medical interventions as involving judgments of value that are widely accepted within a society: The values they introduce are not necessarily those of the patient. Both make the error of separating the grounds for intervention from the factors that affect the quality of life of those, including the unborn, whom the intervention would directly affect, objectivists by trying to eschew value judgments entirely, constructivists by failing to see that social values are only pertinent to the extent that they reflect determinants of the quality of lives.

Testing to see if a fetus bears an allele for Sanfilippo syndrome is justified at bottom because the lives led by children with those alleles are sadly truncated and may diminish the quality of the lives of others, not because Sanfilippo syndrome is a disease or because people in Western society do not value children whose physical, cognitive, emotional, and behavioral development is massively disrupted. Deciding which types of prenatal testing or which types of molecular intervention are acceptable requires us to ask how the tests and interventions would affect the quality of future lives. Northern Indian uses of prenatal technology to reduce the number of daughters are morally repellent if they are inspired by parental contempt for women. But it is not necessary to conjure up moral monsters in the

Punjab. Many of those who go to the new medical centers may do so reluctantly, bowing to the combined pressures of the caste system and their economic situation, or, perhaps, recognizing that the lot of women in their local society is so hard that it would be better not to give birth to girls doomed to be brutalized. Victims of a tradition whose effects they are powerless to undo, they must choose between evils.

Utopian eugenics attempts to mix individual reproductive freedom with education and public discussion about responsible procreation. Our problem has been to understand whether any talk of reproductive responsibility could be more than the expression of our likes and dislikes. Convinced that there is a moral gulf between terminating a pregnancy because the fetus carries alleles for neurofibromatosis and aborting because the fetus is female, we have explored the possibility of grounding the distinction in an objective notion of disease. I believe the exploration has led us to a better perspective, one that acknowledges the need for making special types of value judgments, those that focus on the quality of human lives.

From this perspective, we can begin to see how Modern Civics 2070 might be taught. A different teacher would start with fundamentals. Responsible reproductive decisions, she suggests, involve thought about the qualities of lives. What kind of a life could a child who developed from this fetus have, given what is known about its genotype and the environments that could be provided? What effects would its life have on the quality of other lives that have already taken shape—on the lives of the parents, of their other children, of people in the broader society to which they belong? Sometimes the decisions are easy. When certain allelic combinations occur, the parents know that neurodegeneration will set in during infancy, that the child can be supported for a few years with elaborate care, that there will be no opportunity for the development of a self. By contrast, in affluent societies, although discrimination against women persists, its impact is not so severe as to make it impossible for those who have two X chromosomes to live happy and fulfilling lives. Nor *should* similar opportunities be out of reach for homosexuals, the congenitally obese, or the left-handed. If prospective parents correctly judge that the presence of a genotype for any of these characteristics signals greatly diminished prospects for a life that will go well, then their judgment indicts the social milieu, not the genotype or the trait. Their predicament is exactly that of the Northern Indians who reluctantly abort their daughters to save them from the fate of being female: The moral problem attaches to the social prejudice. The young utopian eugenicists who imbibe the lessons of Modern Civics 2070 should be moved to fight the social and environmental conditions that artificially cramp the quality of lives that might have blossomed.

But there are bound to be many hard cases. Some women, or couples, will have obligations to others—to children already born or to aging parents—which could not be met if a child with great needs demanded their constant attention. Sometimes time and love are not enough, and prospective parents must consider both whether their society can provide the support required to compensate for a combination of alleles and whether they have the right to demand that support. Modern Civics 2070 does not dictate how decisions are to be made in such cases. Instead, the teacher insists on the importance of attending to the quality of future lives, recognizing the constraints and the competing obligations.

Taught in the more adequate way, Modern Civics 2070 presupposes our ability to make judgments about the quality of human lives. That ability, I suggest, is the foundation of a solution to the problem that has hung over the present and previous chapters. To find the moral compass that the enterprise of utopian eugenics needs, we should turn not to the notion of disease but to the concept of the quality of life. As yet, I have introduced that concept in an impressionistic way, relying on our power to make comparisons between lives that unfold very differently. Ultimately, we shall need more refinement, but for the moment we have sufficient clarification: Utopian eugenics envisages a world in which molecular genetics helps people make free and educated reproductive decisions and in which the education is directed toward enhanced understanding of the likely quality of a nascent life.

What our imaginary class takes for granted is the legitimacy of using abortion as a tool for promoting the quality of human lives. The model couple undertake an exercise in comparative population making, choosing abortion when—and only when—they deem the quality of the lives led if the pregnancy is terminated to be superior to the quality of the lives led if the pregnancy is continued. But isn't this to adopt a detached, calculating attitude toward human life, to overlook the moral seriousness of abortion? There is an important challenge here. It is time to meet it.

10

Playing God?

When a fetus is aborted, a life ends. If that fetus carried alleles that would have caused acute pain once sentience had been achieved, or if there would have been unpreventable degeneration, many people would regard the end as merciful. But by no means all. The demonstrators who wave their lurid placards at the entrances to abortion clinics and the religious leaders who inspire them can discern no mercy in the deliberate destruction of a human life, in the murder of a person. Insofar as they are aware of the explosion of molecular knowledge, they must view with dismay a growing arsenal of weapons for wickedness.

Others who protest the coupling of prenatal testing and abortion have very particular concerns. Men and women who have overcome hereditary disabilities to fashion, with considerable courage, rewarding lives for themselves not only fear that eugenics, even utopian eugenics, will fail to appreciate the possibilities for others who bear genotypes like theirs. They believe that the use of abortion as a tool degrades the value of life, even when it is employed by those who are otherwise well intentioned and far-seeing. Troubled by current practices of amniocentesis and selective abortion, and foreseeing a proliferation of prenatal tests, they are daunted by the prospect of a world in which life becomes a commodity, something to be stamped with approval before birth or labeled "defective" and discarded. Future versions of themselves will testify to the failure of quality control. In a world in which human life is debased, will anyone think it worth coping with the debris of "production errors"?

One of the implications of molecular genetics, typically kept in soft focus, is that the development of new tests will multiply the number of abortions. Indeed, although few people would care to say so, that is part of the point. Ultimately, many severely debilitating conditions may be susceptible to treatment or cure—but probably not all, for there will be instances in which the failure of genes to act in early development is critical. Meanwhile intervention on a narrow scale will accompany the possibility of testing, including prenatal testing, on a broad scale. Our only ways of avoiding much suffering will be to terminate pregnancies. Is this to take on powers that we should not have, to play God? Is it to blunt our sensitivity to the value of human life?

The first charge is one of arrogance. Doctors, counselors, and parents presume to question God's design, to correct the Creator's mistakes. Yet even those who adopt a very particular religious perspective, believing that each individual life expresses some divine plan, must recognize that we intervene in nature, on many occasions and in many ways, to promote what we regard as valuable human lives. Stamping out smallpox, or eradicating the AIDS virus, if we could do it, are not usually thought of as violations of God's designs. Most religions do not insist that we take life completely as it comes. So unless a far more radical condemnation of our current practices is required, the wickedness of abortion cannot lie in the fact that we are arrogantly revising the plan of creation by taking a life. Opponents are surely moved by the fact that this is not just *any* life—the life of a virus or a tree, a fish or a rabid dog—but a *human* life. The fault is not arrogance but murder.

Nurses who care daily for children born with severe genetic disorders, who groom them and administer the drugs that keep convulsions more or less under control, who watch the mothers and fathers sitting patiently without hope that their child will respond to their presence, recognize that infectious diseases can sometimes bring a merciful end. Identifying the condition in advance and terminating the pregnancy is even more merciful than waiting—hoping?—for a fortuitous pneumonia and allowing a child to die. Isn't it cruel to demand that a tenuous, controversial interpretation of texts written centuries ago should compel parents to bring into the world a boy who will suffer the compulsive self-mutilations of Lesch-Nyhan syndrome? Advocates of a right to life counter the question from either of two perspectives. One, couched in a secular idiom, replies by insisting that, however well intentioned, the act of abortion still counts as murder, for it involves the deliberate destruction of a person. The other, overtly religious, objects both to the charge that the reading of scripture is debatable and, more fundamentally, to the concentration on physical and mental well-being. What is important to those who oppose abortion on religious grounds is that we are destroying a being which, whatever its cognitive, emotional, or physiological deficiencies, has an immortal soul. The two different perspectives are often entangled, but they should be evaluated separately. I shall begin with the explicitly religious objections to abortion.

Making sense of the doctrine of the soul has challenged philosophers for over two millennia. Yet however we try to understand the relation of the soul to the body, it is hard to fathom how the abortion of fetuses bearing genes for Sanfilippo syndrome or Lesch-Nyhan could be anything other than an act of kindness. The *physical* life of an individual is ended but, presumably, a truly immortal soul would live on, unharmed, perhaps even reunited with its maker. Premature divorce between soul and body has freed the soul from association with a life of physical suffering. Christians often offer consolation for the natural death of a young child after a period of sickness in just these terms: The trials have been endured and the crown is won. If there is to be an objection, then it may stem from the idea that abortion preempts the trials, but only the briefest reflection is required to show that that idea cannot be consistently maintained. Medicine battles against disease precisely to diminish our sufferings: Is it thereby subverting the tests that have been designed for our spiritual progress? Moreover, natural abortions (miscarriages) occur, sometimes as the result of genetic conditions, and apparently these must be viewed as cases in which immortal souls are prematurely, and unfortunately, released.

Utopian eugenics treats abortion as one among many medical techniques for reducing human suffering. Theological opposition to therapeutic abortion should therefore provide some basis for thinking that terminating a pregnancy is an unacceptable means to an otherwise worthy end. What could this be? We are permitted to do what we can to fight disease, to prevent the crippling effects of polio, to use antibiotics to combat bacteria that used to prove fatal. Parents are sometimes encouraged by priests sympathetic to their grief to rejoice over the death of a young child who has been incapacitated by hereditary disease when the death can be conceived as part of God's plan and even to think of miscarriages as reuniting immortal souls with their Creator. But while we can intervene in life, tampering with human lives to prevent or diminish human suffering, we are not to abridge human lives to attain just the same ends, despite the assumed fact that our actions do not touch the most significant aspect of human lives, the immortal soul.

Theologically inspired opponents of abortion surely believe very firmly that we should not second-guess God. They contend that some conditions we encounter in the world, specifically those which involve the beginnings of new human lives, are parts of the divine plan and therefore not to be changed; others, like the assaults of microorganisms or the severe mental retardation produced by the buildup of phenylalanine, are, apparently, not expressions of God's wishes, and thus are legitimate targets for human amendment. Theologians think they can draw the distinction, that they can tell which parts of human suffering are divinely intended and which serve as challenging problems for doctors to solve. We are allowed to do what we can to mitigate the excruciating pains felt by Lesch-Nyhan boys and to prevent them from gnawing at themselves, but we are not permitted to test in advance and forestall the agony that would have been by terminating a pregnancy.

What makes the one response to human suffering appropriate, but not the other? It seems impossible for us to determine a difference, and theological opposition appears eventually to come to rest in Alexander Pope's response to the problem of evil:

> All discord, harmony misunderstood,
> All partial evil, universal good.

As we struggle to eliminate partial evils through a variety of medical interventions, theologians must believe that they know enough about universal good to set limits—and that the merciful abortion of a malformed fetus lies beyond those limits.

The strident slogans of the anti-abortion movement adopt a simpler, apparently secular approach to the issue: Abortion is murder because the clinical talk of "fetuses" disguises the fact that doctors are destroying a child, a very young person. The question "Is the fetus a person?" has often dominated the abortion controversy, insinuating the idea that the correct answer to this question would completely resolve debate. Many opponents of abortion, lured on by the prospect of a quick argument for conclusions that accord with their deepest convictions, offer an affirmative answer without thinking about what attributing personhood to a fetus might entail. It is enough that there is an apparently forced checkmate: Since the fetus is a person and it is wrong to kill a person, abortion is morally impermissible. On the other side, advocates of women's rights to choose abortion often believe that the issue is clinched by giving a negative answer. Denying that fetuses are persons, they can conclude that women have the moral right to do what they wish with the contents of their wombs.

Both kinds of popular rationale have oversimplified the question. Even those who think that fetuses are persons sometimes maintain that it is permissible for a woman to undergo abortion if her life is threatened—and, to dramatize the case, we may suppose that the fetus also bears a Tay-Sachs allele (so that, even if the pregnancy were to proceed to term, the infant would suffer neural degeneration and early death). One person's rights may be overridden in the interests of respecting the rights of another, and some *defenders* of the right to abortion have even conceded the personhood of the fetus, arguing that, in a conflict of rights between mother and fetus, the rights of the woman to determine her own life may take priority over the right to life of the fetus. But talk of conflicts of rights does not fit with many women's experiences of motherhood, for it portrays the fetus as an alien being, an invader of the uterus. More fundamentally, it is hard to understand how invoking the possibility of a conflict of rights is supposed to narrow the scope of the forced checkmate. Even in the extreme case, when the mother's life is threatened by a fetus with a Tay-Sachs allele, it is not obvious why an outsider—the doctor—may kill one to preserve the other. Once it is conceded that the fetus is a person, the doctor is placed in the unenviable position of witnessing a conflict between two people, both with the right to live. Third parties sometimes face the terrible choice of saving just one of two people whose lives would otherwise be lost, but we do not believe that a morally correct response to a mortal combat is to kill one of the two contenders. When the risks to the mother are not so grave, when the consequences would be distress rather than death, admission that the fetus is a person seems to allow no softening of the ban on abortion: Could we be justified in overriding *all* the rights of the fetus, as we would be by terminating its life, to alleviate the mother's distress? To combine the doctrine that the fetus is a person with some tolerance for abortion—for example, in cases of threat to the mother's life—is tacitly to think of fetuses as lesser persons, small, shadowy, hidden persons whose rights are correspondingly scaled down. Although this conception underlies much public discussion of abortion, it is far from clear how to make sense of it.

Nonetheless, it is salutary to think carefully about the speed with which we move from identifying fetuses as persons to concluding that it is wrong to terminate their lives. Sometimes we are inclined— or tempted?—to think that killing is a merciful act, and such situations are most germane to our present concerns. One of the most moving scenes in the film *Schindler's List* shows a doctor and nurse hurriedly preparing and administering lethal injections for Jewish patients too ill to be moved, allowing them to die—indeed terminating their lives—before the storm troopers riddle the ward with bullets. Even if we suppose that a Lesch-Nyhan fetus is a child, a person with rights, perhaps we might see the termination of the pregnancy as an act of kindness, something that will forestall the pains the boy would later inflict on himself, just as the Jewish patients were spared the fear, indignity, and pain of the deaths the Nazis intended. Asserting the personhood of the fetus does not imply, without further argument, that ending a fetal life cannot be morally defensible, even praiseworthy.

Nor does the simple denial that fetuses are persons settle matters as quickly as many defenders of the right to choice appear to believe. Women who have had amniocentesis, who have received bad news about the chromosomes found in cells of the fetus (typically the presence of an extra copy of chromosome 21 that signals Down syndrome), and who have chosen abortion, have testified to the difficulty of the decision and to the anguish they have felt. Abortion cannot be assimilated to trimming nails or cutting hair. The loss of the particular cluster of cells that constitutes a fetus has a moral gravity that should be recognized, and that cries out for a supplement to the doctrine that the fetus is not a person. An adequate explanation of why abortion can sometimes be morally permitted must show why abortion is morally serious even when it is not wrong. For conservatives and liberals alike, answering the question "Is the fetus a person?" should not be taken to settle everything.

Only from a secular perspective should that question matter. Those influenced by religious doctrines should view biological details as irrelevant, since, for them, of primary import is the association between a developing organism and an immortal soul. Yet if the focus is on the eternal, rather than the temporal, the talk of persons and rights should simply drop away, and the questions about whether abortion can serve as the merciful release of a soul from a mangled body should be confronted directly, as they were above.

Once the fetus is considered an organism, and nothing more, thinking of it as a person with rights appears extremely strained. At early stages of its existence, those stages at which abortion is legally permitted, a fetus does not have the neuronal connections that enable it to feel pleasure and pain. Now a minimal condition for having interests and rights is surely that a being have the capacity for pleasure and pain. Nonsentient things do not have rights. An immediate consequence is that, prior to twenty-six weeks' gestation (to take a conservative estimate of the time at which the relevant neural connections form), fetuses are not persons whose rights can be violated by abortion.

The simple argument just rehearsed is likely to be resisted, and for a very obvious reason. Denying fetal rights appears to make abortion morally trivial, leaving nothing to counterbalance maternal interests, desires, even whims. Reflective people abhor this vacuum, and having no alternative way to express their sense of the value of human life, fall back on the doctrine that the fetus is a person, with all of its apparently incoherent ascriptions of rights to the nonsentient. If we had a rival idiom for thinking and talking about abortion, one that would honor the conflicting demands many attempt to satisfy, then the debate about whether fetuses are persons would probably evaporate.

Thanks to the philosopher and legal theorist Ronald Dworkin—whose ideas have inspired both the previous discussions and those that follow—we have an alternative way of conceiving the issue. Dworkin argues, as I have done, that centering the debate on the personhood of fetuses misses fundamental issues. Instead, he suggests, we should recognize that differences about the legitimacy of abortion are traceable to alternative conceptions of the value of human lives. Usually, when a person's life is filled with suffering or is prematurely abridged, our sadness is directed at the consequences for the person, and we overlook the loss of "detached" value, the value of human life, abstractly conceived. When a presentient fetus dies, either naturally or through human intervention, there is no violation of rights or interests because there is no bearer of rights, but something valuable is still lost, prompting grief and regret.

How does value accrue to a life? According to Dworkin, this is the point of serious dispute that generates the abortion controversy. Those who would sharply limit the conditions for abortion or ban abortion outright hold human lives to be valuable—"sacred"—simply in virtue of their conception as human lives. Liberals about abortion take the value of human lives to arise differently, as the result of human investment in lives. In consequence, they view the continuation of some pregnancies as greatly diminishing the value of lives— the lives of parents, children already born, others who are affected—in which there has already been substantial human investment, and see this diminution of value as a far greater loss than the cessation of a human life that has not yet "begun in earnest."

Dworkin assesses the value of lives by looking backward, by considering their origins and their historical development. I shall develop his important ideas slightly differently, by looking forward as well as backward, and by thinking of the value of lives in terms of their past and their potential. When we contemplate a life, we think not just of its origins, of its conception, not just of the nurturing before and after birth, the care that others have given, the ways the person's own efforts have shaped that life's course; we also consider the future, the possibilities that the past efforts have opened up.

Our sense of the tragedy of premature death responds to all these facets of human life. When lives end in infancy or early childhood, our sense of regret at "wasted" love, care, and effort is less, our sense of the human potential is large but unfocused. Death in old age impresses us sometimes with effort rewarded, potential actualized. Most tragic are deaths in adolescence or early adulthood, when the effort of developing, of shaping a person reveals itself in clear possibilities on the verge of realization. The poignancy of a death at any stage is not simply a matter of age nor of the efforts that have been made in forming the foreshortened life: The future potential is also important. Loving care, nurturing, and self-development may produce a person committed to and prepared for some important project, for example, a young man dedicated to tracking down the causes of a deadly disease. Other lives may turn out less happily, despite a comparable history of effort and concern, so that a second young man has no clear project, no obvious opportunities. In the latter case, we might well find it sad that the care and love had not produced a better outcome and might regret the developmental stage at which, inexplicably, things went wrong. But we would not find the second man's early death so tragic as that of the first, the man with "all the world before him."

Premature death leads us to consider the value of the life lost, so that reflections on our sense of tragedy highlight those features that affect the value of lives. Applying these thoughts to the cases of abortion that concern us, cases in which the fetus is found to have a debilitating genetic condition, we can appreciate the sense of anguish and loss that comes *whatever* is done. Before the prenatal test results were in, there was unknown potential: There was no guarantee, of course, that the life would turn out well, that it would either be fruitful or happy, but, equally, there were no reasons to think that the person-to-be could not take the great gamble of developing those individual aspirations that constitute a self. In the wake of the bad news, all horizons are narrowed. Perhaps the fetus bears the genotype for Canavan's disease, so that early neural degeneration and death are inevitable; or maybe the allele for Lesch-Nyhan syndrome is present, ensuring both pain and limited self-development; or possibly what is found is a genetic translocation known to interfere so profoundly with development that the future child, adolescent, adult would linger for years in a hospital bed, unable to engage in anything more than the most rudimentary infantile activities. Because a couple's hopes have been bound up in the thought of a potential person for weeks, there is real anguish at this compression of possibilities to a small sphere of bleak futures as a result of the quirks of meiosis and fertilization. Should the parents continue to give more of their love and their lives to bring into being a sentient person, one who will be vulnerable to pain but who may never be able to form any sense of self? Or should they end this life now, feeling deep regret not only because what they hoped for might so easily have been, but also because a human life, however botched, is ceasing, yet recognizing that there is as yet no person, no subject of pain or pleasure?

When early prenatal tests disclose a devastating genetic condition, we may think of the value of the life that has been created in several ways. Looking backward, we recognize that there has been, so far, rather little human investment in it—much less than in the life of the bright young woman who is killed by a drunken driver as she travels from her home to university. Looking forward, the prospect shows only a sequence of hospital beds, with no chance of autonomous activity. Since continuing the pregnancy to term will narrow the scope of other lives—for, if the pregnancy is continued, the fetus will become a person, requiring and deserving love and attention—that course of action will diminish the quality of the lives that are lived. To end the life before sentience, before there is a person, thus appears most deeply respectful of the value of human life, for it minimizes an inevitable diminution of the value of the lives involved, while avoiding violating the rights of a person.

However, there is a third mode of evaluation, a perspective from which the relative values are assessed quite differently. Opponents of abortion believe any product of human conception to be so valuable that it would be wrong to destroy it, and even though the lives of others may be compressed, the lessening of the value of those lives is insignificant in comparison. But what is the source of the great value that accrues to any human life—to any fetus, irrespective of its genetic constitution—simply in virtue of its origin in the fusion of a human sperm and egg? Tracing the biological processes of fertilization and early development reveals nothing distinctively valuable about *human* lives, nothing that we do not share with our animal relatives. Attending to the way in which the mother and father have charged the biological process with personal meaning, or to the differences that would emerge in the future as the fetus develops into a person, reintroduces modes of evaluation already found to bestow a value on the nascent life that can be overridden by other threatened losses. What remains to support the judgment that this life is so precious that it must be continued?

At first sight, that judgment appears to rest on a truly pathological assessment of the quality of human lives, for it is quite blind to the cramped possibilities for self-development and the sufferings of the future person. But perhaps the thought is that the fetus is a *potential person* and that the lives of potential people are immeasurably valuable. There are surely innumerable potential people whose nonexistence most of us can contemplate without a sense of serious loss. A significant number of children *could* have been formed from Ronald Reagan's sperm and Margaret Thatcher's eggs, or from any of a vast number of equally implausible combinations. Each day and night, couples galore rely on methods of contraception to doom the prospects of any number of potential people. The mere fact that the fetus is a potential person thus fails to invest it with a value sufficient to override other concerns. Yet the fetus grows from a cell in which sperm and egg have already fused, so perhaps it is "closer" to actual personhood than the sperm and egg separated by the diaphragm. Strained metaphysical exercises might offer reasons why this "proximity to actual personhood" matters, why after the DNA from the sperm combines with that of the egg, we have come close enough to actual personhood for the value of the nascent life to overwhelm the pain, the suffering, the degeneration, the early death, the grief that will ensue if that life continues. Few people, friends or foes of the abortion of severely disabled fetuses, are moved by such scholastic niceties. What seems to lurk in the background is an unexpressed religious doctrine, the view that this is not simply a potential person but a life God has planned, a life it is not for us to end, or, perhaps, the doctrine that the fetus already bears an immortal soul. If this is so, then the apparently independent invocation of the value of human life returns again to the cloudy theological issues with which we began and to the difficulty of integrating the religious doctrines into a coherent conception of human life and human suffering.

Abortion decisions are sometimes so complex and multifaceted that it would be folly to pretend that the moral perspective that has emerged, both here and in the two previous chapters, is sufficiently developed to resolve all or even many questions. Nevertheless, focusing on the quality of the lives that people lead and on the sources of the value of presentient human life provides a response to objections that prenatal testing and abortion are always immoral means to otherwise laudable ends. Reproductive responsibility begins by examining the quality of the life that a person who developed from a fetus could be expected to lead and by scrutinizing the consequence that person's birth would have for the quality of the lives of others. Sometimes the genotype present causes such terrible effects, in all environments we know how to contrive, that the quality of life for the possible person who would be born is inevitably extremely low, and the obligations owed were that possibility to become an actual person—obligations that would be discharged willingly—would greatly lessen the quality of other lives. These are the clearest cases, and they have figured in the discussion to plead for the moral permissibility of some abortions.

Far more subtle and problematic examples await. Many hereditary conditions do not doom their bearers to life in a hospital bed. How terrible must the pain or the restrictions on activity and development be to reduce the quality of a life so far that abortion becomes an act of mercy? In thinking about the value of lives, should we simply aim for the largest total value, treating all lives equally, as terms in a colossal sum—assuming that, in doing so, we do not violate the rights of any person? These are hard questions, and answering them will require replacing impressionistic talk of "quality of life" with a more careful analysis. Moreover, there may be no definitive answers to them, simply a recognition that morally responsible decisions may sometimes respond to competing demands in different ways. It is, after all, no part of utopian eugenics to preempt individual decisions. Champions of utopian eugenics believe that moral education is important, that responsible reproductive decisions require informed reflections on the quality of the lives that would be led. They maintain that it is essential to combat social prejudices that would artificially limit the qualities of some lives. In the end, however, they honor the principle of reproductive freedom, recognizing that even in situations where there is no hint of prejudice, morally sensitive people may differ in their assessments of the qualities of potential lives.

For the moment, however, it is important to address a very particular concern. Metaphors come easily in discussing the quality of lives: We talk quite naturally, for example, of "investment" in lives and may proceed from there to view the actualization of potential in terms of "returns on investment." Do we thereby reduce life to a commodity, something whose quality is to be measured, something to be discarded if it does not come up to scratch? And, in consequence, do we erode respect for those born with conditions that are typical targets of prenatal prevention, perhaps even depriving them of the support they need to live satisfying lives?

Babies and children are not commodities, and recent years have brought advances in the rehumanization of gynecological, obstetric, and pediatric practice. Without debasing the processes of pregnancy and birth by making them part of a production line subject to social monitoring, it is possible to recognize that lives differ in quality. A child who will always be confined to a hospital bed will live a life of lower quality than one who can engage in a broad range of activities;

a boy with compulsive tendencies to self-mutilation will experience far more pain than other children; a girl whose mental capacity never proceeds beyond that of a neonate will never have the opportunity to determine her own goals and aspirations. These judgments are very different from the imposition of social values, rightly feared and rightly resisted. Elitist differentiations of alphas and epsilons, that favor those who are athletic, intelligent, good-looking, and well adjusted, all judged by prevailing social standards, offer far too particular and partial a conception of the quality of life, overlooking the fact that people who fare poorly in their ranking would have just as much opportunity for happy and fulfilling lives as their alleged "betters" but for the attitudes that underlie the grading system. Human beings find value in a wide variety of different things, in home, family, and community, in contributing to the lives of others, in achievements of all kinds. What is important is that they should be able to determine what matters to them, and should have a chance to attain their goals. The boy with Lesch-Nyhan and the girl who persists in an infantile state lack any such opportunities. Declaring that their lives have reduced quality does not depend on an elitist judgment that some aspirations are superior to others; it simply shows an appropriate appreciation of their plight.

We are not treating human life as a commodity simply by recognizing that some conditions—those that bring acute pain, confine activity within severe limits, or debar the cognitive-emotional development that allows people to work out what they find significant—violate the preconditions for human lives of genuine quality or by acting sympathetically and mercifully to end such lives before they become the lives of sentient persons. Yet even if our predictive skill continues to grow, as molecular genetics identifies more genes and the relations among genotypes, environments, and phenotypes become ever clearer, there will continue to be people who are confined, racked with pain that prevents any significant action or fixed at very early developmental stages. The wards of hospitals and other institutions will still contain victims of birth trauma, abuse, and accidents.

How will we respond to these people if we have convinced ourselves that the quality of their lives is so low that preventing other lives like theirs is an act of kindness?

Many people who are most at risk for conceiving a child with a genetic disease or disability are reluctant to take prenatal tests to determine if a fetus carries the unlucky genotype, because they have a relative, whom they love, who already has the condition. Knowing in advance would raise a terrible dilemma. Because they are vividly aware of what life with that condition is like, pity would urge them to an act of kindness, to end the pregnancy—and yet they cannot face the betrayal, the devaluation, of a person they love. Parents who have faced the dilemma and who have eventually resolved against abortion sometimes express their feelings in a poignant question: "How could I explain to my daughter (son) that I had had an abortion because of exactly the condition she (he) has?" The fact that the child might be quite incapable of requesting or understanding any explanation does not blunt the force of the question or spare the parent the feeling that abortion would be treachery.

Reconciliation is possible, at least in principle. Contemplating a life in prospect, when there is yet no subject of pain or conscious experience, a couple may judge the quality of that life to be so diminished as to make it right to end it now. Interacting with their child, a person already brought into being, the limitations of the life are already given, and they do what they can to enhance its quality, to show the respect that all persons are due: Indeed, helping actual people with great needs may require *not* bringing into being potential people whose demands would be similarly extensive—recall how the lives of Cypriots with thalassemia have been enhanced by the considerable reduction in the frequency of those born with the disease. The parents recognize and honor the moral demand to do what they can to maximize the quality of their child's life while believing, with anguish and regret, that it would be better for the lives of fetuses with the same genetic condition not to be continued. Abstract morality solves the parental dilemma by distinguishing contexts. Prospectively, before there is yet a person, it is permissible, even praiseworthy, to do what we can to prevent human lives whose quality will inevitably be sadly diminished. Retrospectively, we must respect and love those people whose lives are similarly bounded and act to ensure that their lives have the highest possible quality.

Abstract consistency is all very well—but real parents, confronted with the knowledge that a fetus bears the same genotype present in their severely afflicted son, will surely find it impossible to comfort themselves with fine distinctions or to compartmentalize their reflections so neatly. They need an answer to the question they imagine the boy asking—although they know he is unable to frame any question, unable to understand any words—"Why did you decide to end a life just like mine?" They need to be helped to see that aborting the fetus is not necessarily a rejection of their son. Perhaps they can draw on the abstract distinctions to *begin* a response: "We express our love, concern, and respect for you by doing whatever we can to make your life more comfortable, happier, less confined. Yet we feel our efforts to be inadequate, we wish we could do more. In deciding that you should not have a sister who would suffer as you do, we are not rejecting *you*. Far from it! In thinking about our potential children, the brothers or sisters you might have, we consider the lives we might make possible for them—and the effects that our endeavors for them would have on your life. Just because we are aware how little our efforts have done for you, just because we know that you need our attention and love, we think it would be wrong to have two children for each of whom we could do less."

Endeavors to stop the birth of people who would lead lives of very low quality can be combined with respect and deep emotional attachments to those who are born, expressed in a determination to do all that can be done. So long as prospective and retrospective evaluations are inspired by people and their suffering, human lives will not be treated as manufactured goods.

Yet there is a last version of the worry. Is planning itself antithetical to a proper respect for human life, indicative of a misplaced pru dence that belongs to the world of mercantile calculations? Is the thoroughly planned life worth living? No amount of prenatal testing will guarantee that a couple has children who are assured health and happiness. Because the world is a risky place, having children does indeed give hostages to fortune (in Francis Bacon's famous phrase). Try as they may, parents cannot guard against all the possible angles from which misfortune may strike, but that does not make it inappropriate to try to ward off foreseeable catastrophes. Couples who decide, with distress, to terminate a pregnancy in which the fetus has been diagnosed as bearing the genotype for Canavan's disease are not seeking warranties that they will bear only children whose lives will flourish. They aim to remedy situations in which mischance has already struck before sentient life begins. Instead of a pathological—and impossible—attempt to evade all risks, they act in ways that are comparable to those of loving parents everywhere: Where we see disasters impending, we take steps to prevent them, keeping medicines out of reach, insisting on the use of bicycle helmets, and so forth. There is no illusion that all contingencies can be anticipated and met—unexpected dangers and reversals are always possible—but that hardly means that we should not plan to avoid certain suffering, that care signals disregard for the value of life, or that *im*prudence would express a more appropriate attitude.

Prenatal testing and abortion need not be the apparatus of Faustian scientists who hope to play God; they may be instruments for the prevention of suffering. But perhaps the defense of utopian eugenics has been made too easy by keeping our eyes fixed on the most devastating genetic disorders. The history of eugenics is full of warnings: Once genes step onstage in our reflections about life and death, they expand their roles, recommending genetic "solutions" to problems far removed from those that initially excite our sympathy. Time and again, our predecessors heard the sirens sing that genes are destiny. Can we resist?

11

Fascinating Genetalk

Almost everybody knows how to make some genetalk. Newspapers have no hesitation in mentioning genes on the front page, the barber who does his best with my thinning hair asks politely whether genes for baldness are in my family, even jocular references to genes have become *de rigueur.* Pondering the legal penchant for removing the last ounce of flesh, the British writer Peter Mayle offers today's obvious explanation: "All lawyers have it. They can't help it; it's in the genes . . ." One man's joke is probably another's considered view of humanity. Attributions of various kinds of action to genetic causes are becoming commonplace in popular culture, and those who have "the wrong genes" are consequently doomed. If it's in the genes, it can't be changed.

Those with a license in genetalk know better. Virtually all biologists publicly disavow genetic determinism, insisting that both genes and environments are involved in the formation of phenotypes. Nevertheless, they frequently make reference to genes *for* phenotypic traits—genes for color blindness, genes for colon cancer, even sometimes genes for aggression and genes for homosexuality. Naively, we might think of human bodies (or other organisms) as built up through the individual efforts of an army of DNA laborers, each making an isolated contribution: Here the "gene for eye color" goes to work injecting its special pigment, there the "gene for muscles" assembles a host of cells. Even the development of the simplest organisms, or the simplest structures in complex organisms, belies this innocent picture, involving interactions among gene products and absorption of molecules from the environment that help to trigger the activity of genes. All this is banal for geneticists, but obscured by their convenient shorthand.

If we are indeed launched on a new venture in eugenics—utopian eugenics—then we should pay careful attention to the mistakes of the past. Besides the tragic history of violating reproductive freedom, the compulsory sterilizations and the Nazi genocide, one prominent error has been the assumption that all manner of interesting human traits were under genetic control. Davenport's attempts to find genes "for thalassophilia" in families of naval officers probably did little social harm—but his conviction that some Mediterranean "races" harbored genes "for nomadism" was part of a systematic view of the inferiority of people from certain parts of Europe, ultimately expressed in an immigration policy that returned ordinary human beings (even, surely, some exceptionally talented human beings) to fascism and death. Similarly, the mania for finding genes "for feeblemindedness" not only led to the sterilization of women of average intelligence but also to the stigmatization of their entire families. As we, and our descendants, ponder difficult social problems, casual genetalk provides an easy way to think that there are simple, cheap biological solutions. There is a danger of repeating the injustice visited on Carrie Buck, unless we learn to think carefully in the ways that professional geneticists do about the exact role of the genes in complex human characteristics. This chapter attempts to articulate distinctions utopian eugenics will need to keep constantly in public view if it is to live up to its billing as something quite different from the eugenic ventures of the past.

Except for a very small number of traits of a very few (nonhuman) organisms, virtually all the details of the processes through which phenotypes emerge are unknown. We know the shape of the story—but only as we might know of a mystery novel that it starts with a murder, involves a detective collecting clues and questioning suspects, and concludes with a brilliant unmasking of the culprit. A new human life begins with a zygote, a fertilized egg whose nucleus contains a distinctive collection of DNA molecules and whose cytoplasm is rich in proteins bequeathed by the mother. Without these proteins, the DNA would be inert, incapable of participating in the sequence of events that constitutes development; but because the appropriate enzymes are available, genes are transcribed and translated, forming new proteins. As a result of this internal activity, the cell divides. More genes are switched on, to help in the building of even more proteins. Cells divide further, exchanging molecules that cause them to behave differently, at first in subtle ways, later more dramatically. Because one cell is placed so that it has more neighbors, its concentration of a protein may be increased or reduced, allowing for a special reaction that eventually activates an unprecedented gene; so differentiation begins. Now sorted into types marked with special molecular milieux, the cells continue to divide, interacting with one another and with the environment of the developing fetus. New molecules flow down through the placenta, to be absorbed and to play their part in the intracellular choreography. Eventually, out of billions and billions of local reactions, there emerges a baby with identifiable characteristics. Outside the womb, the molecular stimulation continues in a wider environment, as the child's growth is shaped by the air breathed, the food consumed, and in ways that are surely as profound as they are currently inscrutable, by the bombardment of ear, eye, and skin which sculpts the wonderful architecture of the brain. In the end, an adult experience—be it pride in a child's success or sympathy at a friend's loss—has its origin in innumerable small physical and chemical reactions in which many genes and many environmental factors have played small but critical parts.

Despite all the complexity, there are sometimes identifiable regularities. If a mutant allele is present at a locus, yielding an atypical protein, there may be little disruption of development apart from a single distinctive effect. Recall the example of sickle-cell anemia. The disease results when a person bears two copies of a mutant allele at the locus that directs the formation of one of the globin chains that occur in adult hemoglobin. A single amino-acid substitution— glutamic acid for valine—seems to have no impact on human development except to produce *erythrocytes* (red blood cells) containing atypical hemoglobin. Unluckily, that small difference is enough to cause episodes of acute pain, collapse, and premature death. When the attacks will come and how severe they will be depend on features of the environment that are currently unfathomed. The environmental variation drapes an apparent genetic-developmental regularity: People with two mutant hemoglobin alleles have erythrocytes that assume relatively rigid sickle shapes under conditions of low oxygen; the variable effects on health and longevity can be attributed to differences in individual physiology and to the environments in which those who are afflicted live.

Huntington's disease provides an even purer, equally tragic, example. Apparently, people who have an over-long CAG repeat near the tip of chromosome 4 will develop the disease no matter how they live. Sufferers have quite different genetic backgrounds; they grow up and spend their lives as adults in heterogeneous ways. None of this seems to make any difference. The sad termination of their lives is fixed by the presence of the extra CAG codons. The disease, we are tempted to say, is determined by the genes.

Genetic determinism has its intellectual roots in cases like this. Because we do not understand the mechanism, we can only look at the termini—the genotype and the phenotype. Observing that the phenotype comes about independently of the genes present at other loci, independently of the environment, we pick out the long repeat as the crucial causal factor. The body of the Huntington's patient, including the brain that will be the site of the dreadful decay, is built up out of a vast number of gene-environment interactions, and yet one feature—the susceptibility to neural degeneration—seems to depend not at all on the details of the environmental shaping, or indeed on anything other than a sequence of bases near the tip of chromosome 4.

To a first approximation, we could identify those characteristics of people that are genetically determined by picking out traits that always develop when a particular combination of alleles occurs, no matter what the environment in which the person develops and lives, no matter what the genetic background. This is only an approximation, because some genes at other loci or some environments might make it impossible for the characteristic to appear. Quite probably, nobody has ever been unlucky enough to bear both the Huntington's long repeat and two copies of the Tay-Sachs allele, but, if there were any such people, they would die in childhood, before the onset of Huntington's disease. Enough people have died in wars and epidemics in our century to make it almost certain that some among them have carried the long CAG repeat. Of course, none of this is at all comforting, and shows only that environments or genetic backgrounds that prematurely terminate a person's life may mask the phenotypic expression of the Huntington's allele. Far more importantly, the search for ways to prevent and treat the disease reflects the hope that there are *presently unknown* environments—possibly involving as yet unanticipated modifications of bodily processes—in which people bearing the long repeat would not suffer the disease. Medical researchers seek environments in which people live and thrive beyond the time at which the disease would normally appear, and the special interest in environments that permit human flourishing should be explicitly recognized in our understanding of genetic determinism.

Our genetically determined characteristics are those that would persist under those changes in genetic background and in environment that allowed us to survive and thrive. At present, Huntington's disease *appears* genetically determined, for we do not know of any ways in which to modify environments to allow for the continuation of healthy life. But researchers aim to show that what is genetically determined is not the disease, but something else from which the disease can be separated by ingenious environmental changes. With luck, our descendants will rechristen the long CAG repeat, viewing it as fixing some molecular condition that, *if left untreated,* would produce neural degeneration.

Our current attitude toward PKU prefigures this optimistic scenario. Before the discovery of special diets that enable children to develop normally, it was natural to think of a disease most prominently revealed in severe mental retardation as genetically determined. Armed with the understanding that the immediate causes of the cognitive disability lie in overloads of phenylalanine and undersupply of tyrosine, we can separate the manifested disease from the underlying genes. What is genetically determined, after all, is an inability to break down phenylalanine in the normal way, and it is possible to compensate, although not perfectly, by creating an environment in which phenylalanine intake is drastically reduced.

Nobody yet knows how to achieve the same for Huntington's disease, nor can we be sure that it will be possible to separate the genes from the phenotypic trait—the neural degeneration—we seek to prevent. Sickle-cell anemia lies between PKU and Huntington's. Because of the wide variation in frequency and severity of the crises associated with the disease, it is clear that there must be differences among individuals, traceable to combinations of environmental factors and background genes, that account either for a variable tendency of cells to sickle or for heterogeneous effects of sickling. Sadly, however, they are as yet unidentified. Were we to succeed in singling them out, we should no longer think of the *disease* as genetically determined: Instead, we would pinpoint some more fundamental molecular condition as the inexorable effect of the genes.

There is a first moral here. Casual claims of genetic determinism typically propose that some exciting human characteristic lies within the genes' iron grip. Moving beyond the simple association of traits with alleles, across a more or less broad range of environments, we find, as we probe the mechanisms, that if something is genetically fixed, it is not quite what we thought. The likely candidates for genetically determined properties are conditions describable only in a complex molecular idiom, conditions that translate into the traits that fascinate us only in the presence of background causal factors. Peter Mayle's ravenous lawyers may share genes for something, but I am prepared to bet several kilos of truffles that it will be a neuro-molecular state which issues in Shylockian demands only under the pressure of quite special social forces.

Our eugenic past resounds with declarations about the genetic determination of human behavior. If legitimate fears about eugenic revivals are to be allayed, then it will be important to assess the impact of molecular genetics on our understanding of human behavior. Clarifying conceptions of genetic determinism is a first step. But the corruption of genetalk is so pervasive that we should pursue further efforts in intellectual hygiene.

Are there genes "for" homosexuality, aggression, or low IQ? Better to postpone the question, asking first in what sense there are genes "for" anything. Classical and molecular geneticists have used the locution for decades: Thomas Morgan's great breakthrough in the study of heredity began with the observation of a mutant fruit fly that carried the gene "for" white eyes; he and his students quickly discovered dozens of genes for eye color, each identified through discovering flies whose eyes differed from the normal shade of red. Here lie the roots of contemporary conceptions. Geneticists discover genes "for" particular traits by showing that genetic *variation* at a locus is associated with *changes* in the trait. The association is typically recognized in organisms whose genetic background appears otherwise normal and who inhabit environments that are standard for the species.

Because of the complexities of development, it is helpful to start cleaning up genetalk by making the dependency on other genes and on the environment explicit. Relative to a background constellation of alleles at other loci and to an environment (or more exactly, to a complex sequence of environments in which the organism develops), a locus affects a characteristic when, as we find or create different combinations of alleles at that locus, organisms with the appropriate genetic background, developing in the pertinent environment, show changes in the form of the characteristic. So geneticists breed vast numbers of flies with the same combinations of alleles at virtually all their loci but with genetic variation at a single locus. They rear those flies in standard laboratory environments and show that the different combinations of alleles at the chosen locus are expressed in variations in eye color. They conclude, therefore, that that locus is "for" eye color and that the alleles found there are forms of the gene "for" eye color.

Pedantry is important here. Particular alleles at the locus, either singly or in pairs, can be said to be "for" the special form of the trait with which they are associated—*relative to the genetic background and the environment.* But if we are going to talk about genes "for" traits without any qualification, we have to have in the back of our minds some idea of *standard* genetic backgrounds and *standard* environments, relative to which the locus affects the trait.

How should we decide which background genes and which environments are standard? Classical and molecular geneticists typically operate with intuitive notions of what counts as a "normal genotype" and a "typical environment." These notions are perfectly adequate for the purposes of everyday genetic research—no confusions or controversies are likely to arise in deciding on "normal" genetic backgrounds for flies or on "standard" laboratory environments—but they become problematic when they are transferred unthinkingly to the human case. To understand what is meant by talking about genes "for" the human characteristics that most fascinate us—intelligence, personality traits, tendencies to sex and violence—it seems appropriate to think of standard genetic backgrounds as containing combinations of relatively common alleles (although not necessarily relatively common combinations of alleles!), but the notion of a "standard environment" is far more bothersome.

One approach would be to think of standard environments as those common to a society's practices of rearing children: There will inevitably be considerable variation across societies and across epochs. Another would take standard environments to be all those that permit a growing child to develop in accordance with species-typical patterns. Yet a third would count environments that allow people to thrive in the ways of most concern to them. The differences among these conceptions are significant. Suppose a scientist succeeds in showing that there is a gene "for" aggression in *all* standard environments of the first type, that is when children are raised in a society's traditional ways. It by no means follows that the bearers of these genes are doomed to be lifelong victims of irresistible urges to lash out, for there may be environments, standard in the other senses, that lie beyond the society's current horizons and permit a more benign outcome. Similarly, it is not hard to see that there can be environments that enable people to attain the things they value most but which do not permit the species-typical pattern of development.

Faced with the suggestion that a gene "for" some fascinating human trait has been found, the canny consumer should ask a number of questions: What sense of standard environment is being invoked? In what range of standard environments—some? many? all?—does the trait invariably occur in those who bear the gene (and have a standard genetic background)? People who are quick to pose these questions will avoid the most prominent fallacies into which genetalk often sweeps us. They will realize that there can be genes "for" forms of behavior that are brought about in complicated ways. For example, there might be alleles "for" dyslexia, meaning that there is a locus at which mutant alleles sometimes occur, so that when someone has a pair of these alleles, an orthodox complement of other genes, and grows up in many of the typical ways in which children are raised in affluent societies, that person will come to have the reading and spelling problems characteristic of dyslexia. (The normal alleles at this locus might even come to be known as "reading alleles.")

None of this violates the obvious point that developing a capacity to read is an extraordinarily complex business, involving all sorts of genes and interactions with the environment. Nor does it mean that dyslexia is genetically determined. Reading and spelling problems may crop up if the child with the alleles "for" dyslexia is reared in most of the traditional ways, receiving instruction of the usual types. But the trouble might be avoided by any of a number of approaches to child development. Perhaps there are a few minor modifications of traditional teaching methods that would do the trick (environments standard in the first sense); or there may be some nontraditional instructional programs that break quite radically with usual ways of teaching children but allow for the usual pattern of development (environments standard in the second sense); or perhaps, even more radically, there may be methods of rearing children that alter the patterns of cognitive development to which we are accustomed without interfering with other desirable aspects of the child's growth (environments standard in the third sense).

Furthermore, there can be genes "for" phenotypic traits despite the fact that most people who have the trait do not have those genes. Arteries become overloaded with LDL cholesterol as the result of different causes. Some unfortunate people carry a combination of alleles that, in all the environments we naturally inhabit, cause them to have livers that are unable to break down the "bad" cholesterol. At the other extreme are dedicated gourmands who assault their systems daily with mounds of fettucine Alfredo. On a more moderate diet, the genes they carry would be quite compatible with normal amounts of LDL cholesterol, and the high levels they actually display testify only to their relentless overindulgence. Between the poles are people whose elevated cholesterol measurements stem from many different sorts of interactions between their genes and their ways of life. So there can be genes "for" high cholesterol, even though the overwhelming majority of cases have nothing to do with those genes and many may be caused by purely environmental factors.

Drawing distinctions is dreary work, but the consequences of conflating them are worse. How many nongeneticists, reading newspaper reports of the discovery of a new gene "for" a trait that interests us have not thought, *So now they know the cause of that!* or *How terrible that people with the gene are doomed to be like that!* or *You can't get that unless you have the gene for it?* None of these responses is justified; all depend on making one of the muddles that exploit the ambiguities of genetalk. Yet reaching these conclusions can have profound effects on our thinking, leading us to abandon environmental interventions that might be very effective. Instead, we gloomily reflect that it's all in the genes and wrongly convince ourselves that nothing can be done to set things right.

Examples of confusion are already with us, and the accelerating pace of identifying human genes threatens to swamp us with more. In the wake of the discovery that a human gene is very similar to a gene whose mutations cause obesity in mice, newspapers and magazines showered the public with "implications": Fat people should not be blamed for overeating; battling the bulge is hopeless; we can expect hormonal therapies to replace diets; we must guard against selective abortion of those with mutations indicating obesity. These claims, which contain elements of insight, were advanced before the pertinent alleles in our species had been sequenced, before the mutations were known, before the percentage of fat people with mutant alleles at this locus had been investigated, and before the frequency of obesity in those who carry such alleles had been measured. In short, those who made the claims could only guess how many obese people had the guilty alleles and how many of those with the alleles were fat. Apparently the message heard by the media was that certain alleles account for obesity, so that all obese people have those alleles and all people who have the alleles are obese. Any number of everyday observations are at odds with this simplistic picture. The increased rates of obesity in lower income groups is surely to be understood more in terms of the kinds of foods eaten by people in those groups than by the distribution of DNA sequences. Moreover, the tentative hypothesis about the mechanism of gene action—that the normal protein serves as a satiety signal—undercuts the naive view that people with mutant alleles are doomed to be fat. Mice who do not receive the appropriate signals may puff up in dramatic ways, but human beings, even young children, have alternative ways of monitoring how their food intake is affecting their physique. To obtain real benefits from the discovery of human "genes for obesity," and to avoid some of the potential problems, we need to dissolve the confusions of casual genetalk.

If our talk of genes "for" traits is troublesome, our ideas about a "genetic basis" for traits or about "genetic disease" can be even more confused. The venerable controversy about nature and nurture is a graveyard of rotting doctrines that are periodically revived by those with no nose for distinctions. Intelligence, aggressive tendencies, schizophrenia, sexual preference—all these fascinating traits are supposed to have a "genetic basis." There is one sense in which this is almost trivially true. However, ardent revivalists rarely stop to ponder just what kind of "genetic basis" is revealed in the data that inspire them. Rather, they move on to parlay a commonplace into something much more exciting, something far beyond the evidence.

To show a "genetic basis," you demonstrate that people who grew up in the same environment either have or lack a trait—either become schizophrenic or do not, for example—depending on which kind of genotype they have. Typically the environments considered fall into quite a narrow range, for that is unfortunately the way of studies of complicated human traits, and most people would feel it unjustified to claim *straight off* that the characteristic of interest is determined by the genes. However, genetalk lets the public get there in easy stages: First point out that there is a "genetic basis," then conclude that there are genes "for" the trait, and finally, announce triumphantly that genes are destiny.

We have already exposed the fallacy of the last step, but the first stage is equally misguided. What is actually known is that, in some narrow range of environments, having one genotype makes people more likely to be schizophrenic (for example) than does another genotype. It is quite consistent with this to suppose that in *other* environments the relationships are reversed or there is no difference. Suppose that there is indeed no difference, and consider Andy and Mandy, and Maud and Claud. Andy and Mandy are brother and sister, as are Claud and Maud. The boys, Andy and Claud, were brought up in very similar ways, and Andy became schizophrenic, while Claud did not. The girls, Maud and Mandy, were also reared very much alike, but quite differently from the boys, and neither Maud nor Mandy became schizophrenic (they are as healthy as Claud). Now the siblings are genetically similar at whatever loci you take to be relevant: That is, Andy and Mandy share the same allelic combinations at these loci, and Maud and Claud are also alike in their DNA. Does this scenario reveal a "genetic basis" for schizophrenia?

In one sense, the answer is clearly "yes": Andy and Claud share a common environment, and the difference between their genotypes makes a difference to their mental well-being. Because of Claud's combination of alleles, he is healthy, while Andy suffers schizophrenia. Yet we might equally point to an "environmental basis" for schizophrenia. After all, Mandy has just the combination of alleles that spells trouble for Andy, but the environmental difference between her rearing and his makes just the same difference: She is as healthy as Claud and Maud. To declare the nature-nurture controversy resolved—in favor of nature—would be wrong. Both nature and nurture play a part in the causal process, and whether we focus on nature or on nurture as the crucial causal factor depends on whether we are concentrating on the difference between Andy and Claud or the difference between Andy and Mandy.

The big temptation, of course, is to think that showing that genetic differences make the difference in *some* environments amounts to demonstrating that they make the difference in *all standard* environments, moving on to the conclusion that genes are therefore destiny. Neither you nor anyone else would commit that fallacy if you thought things through using the distinctions I have drawn—but shorthand is convenient and seductive.

Announcing a "genetic basis" for a trait is one way to fall; another is to declare a "genetic disease." The simplest approach takes a genetic disease to be one that appears determined by the presence of a particular combination of alleles. In all the environments we know how to contrive, people with the unfortunate genotype always contract the condition. Huntington's disease is a genetic disease in just this sense, but that does not mean that the genes inevitably spell doom, for we may hope to find environments (standard in either our second or third senses, nonstandard in our first sense) in which people with the long CAG repeat avoid Huntington's and continue to thrive. Furthermore, a disease can be "genetic" even though typical instances are caused by environmental factors that would override virtually all differences in the genes. There may be diseases that are genetic in this strong sense, hypercholesterolemia for example, for which *some* people who have the disease have genes that produce it in any of the environments we know how to arrange, while the majority of those afflicted have high cholesterol levels because of the foods they eat.

A weaker notion of "genetic disease" would count anything for which there is a "genetic basis." Schizophrenia surely qualifies as a genetic disease, simply because in some environments differences in genotype are systematically associated with having or not having schizophrenia. This conception is so liberal that it may embrace all the ills human flesh is heir to—even colds will prove genetic if there are environments in which particular genotypes make some resistant and others not, this despite the fact that what causes colds are viruses.

Besides "genetic diseases" there are "genetic susceptibilities to disease." Parallel to the more demanding notion of "genetic disease"—that illustrated by Huntington's disease—a "genetic susceptibility" for a disease is a genotype whose possession raises the chance, in all the environments we know how to differentiate, that the bearer will develop the disease. If you have a genetic susceptibility for heart disease, then no matter what advice your doctor gives you and no matter how assiduously you follow it, your risks of heart disease will be greater than those of your luckier friends. As with the discussion of the last paragraph, it is possible to introduce a more catholic conception, letting a genetic susceptibility be any increased probability of disease for which there is a "genetic basis."

To make matters yet more complicated, there is a quite different conception employed in the recognition of cancer as a "genetic disease." Cancerous tumors occur because somatic cell division sometimes produces descendant cells with modified genetic material, and these cells fail to regulate their growth properly. So cancers are very special kinds of genetic diseases, genetic in the rough, intuitive sense that they have something to do with the DNA; more precisely, they are diseases in which the special characteristics of genes in some somatic cells prevent those cells from being able to function normally. Many genetic diseases of this type are not "genetic diseases" of the sort exemplified by Huntington's disease. Cancer, for example, frequently occurs because people are exposed to environmental agents like radiation, cigarette smoke, and toxic chemicals. However, just to confuse matters, some cases of cancer are "genetic" in *both* senses: Not only does the tumor come about because the DNA fails to regulate cell growth and division, but this failure is traceable to the person's native genetic endowment, which would have produced cells with the misbehaving genes in any of the environments we know how to arrange.

Molecular geneticists usually steer with ease through the shoals and reefs of genetalk. The distinctions I have drawn will enable outsiders to avoid the most obvious forms of shipwreck, but even if the dry details are forgotten, there are principles for safe navigation. Announcements about "underlying genes," however they are couched, try to identify important causal determinants of complicated processes, and it is always worth inquiring how the interaction between genes and environment is supposed to go. Are background genes and environmental factors so irrelevant that the effect would be produced anyway, so long as the focal alleles are present? How is the irrelevance of environmental variation demonstrated? With these questions in mind, we can come at last to debates about the genetics of human behavior.

Life has not been kind to human behavioral geneticists. Their colleagues who study flies and worms can bring together the combinations of genes whose effects they want to explore, and they can modify environments almost at will. Researchers on the genetics of human physiology have a harder time, but they can be guided by comparisons with animal models, from which ideas about the impact of the environment can be gleaned. Using animal models to illuminate complicated pieces of human *behavior,* on the other hand, is highly controversial: Even though "forced copulation" occurs in scorpion flies and mallards, we should be suspicious about the idea that probing the genetic basis of such behavior will teach us much about human rape, for the neural mechanisms—to say nothing of the potential social influences—are likely to vary enormously between ducks (or insects) and people. To disentangle the roles of genes and environments in producing human behavior, it is necessary to fasten on "natural experiments," cases in which people sharing genes of interest are brought up in different environments. Pairs of monozygotic (identical) twins reared apart have been the ideal windfalls.

Unfortunately, although these windfalls can be collected by those prepared to take pains, the careful behavioral geneticist who would like to find out whether twins who have lived their lives in very dissimilar circumstances share some interesting trait must frequently draw conclusions from small samples. Even when twins have been separated from birth, they are often reared in ways that are quite alike. Researchers are thus forced to choose between a large sample in which the conclusions might be confounded by a common environmental factor or a much smaller body of data in which the environmental differences are marked.

Molecular genetics has already begun to open up new vistas. As more refined genetic maps become available, it is becoming possible to use genetic markers to track behavior genes through families, and ultimately to pursue the strategy of positional cloning to identify and sequence candidate alleles. With the ability to discover just who carries an allele thought to be implicated in a particular kind of human behavior, geneticists may investigate the extent to which the behavior of people with the allele varies with differences in the environments in which those people develop. Perhaps monozygotic twins will become redundant (although the fact that they share *all* their genes, in the same order, may keep them in business for exploring cases that involve many loci).

A prominent indicator of future possibilities comes from a much-discussed study. Dean Hamer and his colleagues have discovered an association between the forms of a marker in a region near one tip of the X chromosome and male homosexuality. As yet, they have not isolated a gene or explored mutations: The pertinent locus lies somewhere in five million base pairs, in a region that may contain as many as two hundred genes. Even geneticists who have been skeptical about human behavioral genetics have lauded Hamer's work, and with reason. For Hamer is admirably cautious in acknowledging what he has not yet shown. He points out that he cannot estimate the frequency with which those who bear the "gay" alleles develop same-sex preference and that the fraction of homosexuals with such alleles could be at most two thirds, and quite possibly very low indeed ("only a few percent"). However, Hamer's work points the way to new investigations in the genetics of human behavior. Knowing the sequence of an allele that is associated with same-sex preference in a particular family, geneticists could try to discover whether other people with the sequence, bearing different background alleles and living in very different environments, were also homosexual. Assuming (a large assumption) that individuals from a representative cross-section of a diverse society would cooperate, geneticists could obtain a much clearer picture of how the particular sequence of interest affects sexual preference.

To summarize: Classical behavioral geneticists look at the relation between inputs and outputs—genotypes and phenotypes—not at the mechanisms through which genotypes and environments (including the initial molecular milieu of the fertilized egg) produce phenotypes. One great handicap to understanding the relationship of genes and environments to complex human traits has been to discern when two people share the same genes, that is, when the inputs are the same. Relatives, especially close relatives, are so important to behavioral genetics because relatives share genes, and the very closest relatives, monozygotic twins, share all the genes. Gene sequencing gives us the ability to discern common genes even in unrelated individuals, and thus to explore the outputs associated with the same inputs when the environment changes.

But there is a second handicap from which molecular genetics alone will not liberate us. Human behavior, we believe, is influenced by all kinds of processes that we do not know how to describe at the molecular level—by the feelings that accompany particular experiences, the words that are spoken, and so forth. Abysmally ignorant of the basic capacities that underlie our cognitive and emotional development, we cannot have firm ideas about what kinds of changes in the environment would be significant. Our exploration of environmental variation proceeds blindly, and when all the data about sex preferences and sequences are in, we shall still be unsure whether we have demonstrated that an effect is invariant under changes in all the crucial environmental variables or whether we have simply read a very large number of copies of the same newspaper.

Mapping and sequencing are valuable tools, but even in overcoming the first handicap, there will still be plenty of technical challenges. Positional cloning (the strategy of using maps to track down genes, so successfully pursued in understanding cystic fibrosis and Huntington's disease) works most smoothly when geneticists are hunting a single gene. When many genes affect a phenotype, as they do in the behavioral examples of chief interest, markers that are tightly linked to one locus may have no relation to another crucial locus, so that it becomes hard to distinguish between genetic heterogeneity (different instances of the phenotype are brought about in diverse ways) and simply a very weak correlation between a marker and a locus. So far, analysis of polygenic physiological conditions has proved difficult enough, although researchers have been able to isolate loci at which particular mutations exert large effects, as in the study of early-onset breast cancer. Hamer's work may show how similar successes can be achieved for sexual preference and complex psychological and behavioral conditions.

When scientists are concerned with psychological or behavioral characteristics, there will be two further sources of noise: One lies in the vagaries of diagnosis, the other in the possibility that unrecognized environmental causes will confound our judgments. Using genetic and physical maps to track down a gene responsible for schizophrenia, for example, is vulnerable to errors made in deciding who is afflicted, as well as to failure to recognize that some people who lack the gene have other alleles and live in environments that jointly produce schizophrenia. The history of recent searches for genetic bases for various behavioral disorders, in which claims give way to retractions and qualifications, illustrates how serious these problems are. Following the chronicle for schizophrenia, for bipolar affective syndrome ("manic depression"), and for Alzheimer's disease in successive editions of the human geneticist's bible, *On Mendelian Inheritance in Man,* offers a revealing contrast to the great success stories of human molecular genetics. Where we find claims to the effect that a genetic culprit has been found on a particular chromosome, they are tentative and hedged, and with reason. For this year's judgment is likely to be questioned by next year's finding.

To paraphrase that great Victorian counselor, Mrs. Beeton, first catch your gene. Yet even without a technique that assures success, behavioral geneticists may be fortunate, identifying, cloning, and sequencing alleles whose bearers among the families they study have a much higher frequency of some interesting trait—a tendency to violence, or same-sex preference, or schizophrenia. At this point, molecular genetics could really prove liberating in the way described earlier by allowing a systematic exploration of the characteristics of people who carry the sequence and who, superficially at least, inhabit quite different environments. Although there have been advances in uncovering a possible "genetic basis" for homosexual preference, accompanied both by professional restraint and predictable public overinterpretation, much systematic work (identifying the locus and the mutations, tracking the phenotypic effect of the mutations as the environment varies) remains to be done. What would it take to show that homosexuals "can't help it," that "it's in the genes"?

Among male monozygotic twins—who share all the same genes and most of whom are reared together (and thus share very similar environments)—the concordance rate for homosexuality is around 50 percent (for whatever reason, lesbian twins have not so far received the same level of attention). In other words, about half those males who are monozygotic twins and homosexual have a twin who is also homosexual. Although this result, together with the finding that brothers of homosexuals are more likely to be homosexual than are unrelated men—in Hamer's study 13.5 percent of the gay men had a gay brother—is often taken to reveal the "genetic basis" of homosexuality, it would be equally pertinent to point out an apparently very large environmental effect. For all the shared genes and the extremely similar environments—monozygotic twins reared together have a far more similar upbringing than ordinary brothers—about half the co-twins of homosexuals are not themselves homosexual. Dangers of misreporting aside, there must be subtle differences within environments that affect the character of sexual preference quite profoundly. Comparing the class of monozygotic twins to the class of brothers, the high concordance rates reveal the "genetic basis." Comparing the discordant twins to one another, we have to recognize an "environmental basis."

Perhaps within a few years there will be known sequences, assigned to known chromosomal locations—at the tip of the X chromosome, for example—which are associated in some families with male homosexuality. Even before investigators go out to explore variation in sexual preference among men who carry these sequences, they will know that either there are many male homosexuals who do not carry the sequences or that there is a significant range of environments in which having some of these sequences does not issue in male homosexuality. For, suppose almost all male homosexuals have a particular sequence. Suppose also that men who have that sequence are almost invariably homosexual, irrespective of the environment in which they grow up. Given these assumptions, almost all the male homosexuals who also have monozygotic twins would have the sequence. Their twins would also have the sequence. So virtually all the pairs of twins would show the same sexual preference. However, we know that only half the identical twin brothers of male homosexuals are themselves homosexual. Therefore, one of the initial suppositions has to be wrong: Either the sequence does not account for more than half the cases of male homosexual twins or having that sequence does not yield homosexual preference in most environments.

However, even the genetic determinist's most striking data deserve scrutiny. How ought we to react to a painstaking study demonstrating that a particular combination of sequences distributed on different chromosomes is invariably associated with male homosexuality across many environments—when boys are reared by two parents or by one, when they have elder siblings and when they do not, when they are treated gently, harshly, or abusively, when they are neglected, and so on—through the whole canon of environmental features that we might antecedently consider relevant? One possibility is that the genetic determinist account is correct, and that the development of sexual orientation is crucially fixed by the character of a small number of proteins. But this is not the only story: Perhaps there is some hitherto unnoticed common feature of all the environments in which people live—as the fact that our usual diets contain large amounts of phenylalanine was once unremarked—which interacts with some effect of the crucial proteins to shape sexual preference.

Nobody knows very much about the development of sexual preference. Fantasies are healthy if they lead us to appreciate alternatives we might have missed. So let us speculate. First, we can imagine a fanciful determinist scenario. Because of the presence of different DNA sequences at the relevant loci, different proteins are formed, and these different proteins play a role in the setting up of some neural connections. The variations in neural wiring occur independently of signals from the environment, and sexual orientation is simply a matter of the type of neural wiring that a person has (perhaps, as has recently been suggested, it is a consequence of the size of particular clusters of cells in the hypothalamus, a region of the brain known to affect sexual activity in primates). The second scenario agrees with the first almost to the final step: Differences in the proteins translate into differences in neural architecture, but the primary effect of the neural variation consists in a modified response to certain kinds of odors. Imagine that we discover that there is a certain period during which children are very sensitive to the distinctive smells of men and women, and that the association of women with pleasant or unpleasant smells is very important to the way in which boys develop their sexual preferences. Boys with the special sequences develop a neural configuration that is overloaded by olfactory bombardment, at a time at which contact with caregivers is arousing infantile sexual feelings, and the particular molecules emitted by almost all women thus cause sensations of pain and an inhibition of sexual desires toward women.

Both stories are far-fetched, and the truth is certainly immensely more complicated, almost certainly beyond our present imaginings. We can use the second scenario to focus a general possibility, the possibility that alleged genes "for" male homosexuality might more directly affect something quite different, something more mundane that has effects only on sexual orientation because of pervasive but contingent features of our child-rearing practices. Once we discovered that life follows the second scenario, our vision of sexual orientation would change, bringing some liberating consequences as well as disturbing questions.

Male homosexuality, it would turn out, is not genetically determined. What would be genetically determined would be a propensity to be overstimulated by certain odors. The pertinent genes could then be more accurately conceived as genes "for" smell sensitivity. Homosexuals would be revealed as akin to those people who are overpowered by the scent of orchids or sickened by lilies—and, presumably, no more to be despised, spurned, or vilified. But surely there would be a downside. Homosexuals and heterosexuals alike typically think of their sexual orientation as something very important to their identities. Sensitivity to smell seems far too trivial to reach so far into the constitution of the self, and besides, the development of sexual preference would become easily malleable. Mothers determined to rear only heterosexual sons could perhaps administer drugs to dampen olfactory stimulation, or could use special soaps and perfumes during the critical period. By the same token, homosexual sons could be assured by making sure that boys were exposed to sufficiently powerful doses of the crucial scents.

Unless social attitudes change profoundly, there is no doubt about the direction in which the shaping of sexual preference would go. Would it be wrong for mothers to take the relevant steps? After all, once we recognize the underlying genes as "for" smell sensitivity, it seems illegitimate to proclaim that mothers who mask natural odors are tampering with the boys' identities. Because of their genetic endowment, the boys have an unusual sensitivity to particular scents, but the sexual preference is thrust upon them by the contingencies of the childhood environment. Protecting them from olfactory overload will enhance their chances of leading happier lives, for they will no longer have to resist a dominant culture. But there surely is the problem. The judgment that modifying the childhood environment counts as a form of "protection" carries the residue of centuries of conceiving homosexuality as immoral, decades of thinking that homosexuals are sick. Future molecular biology may provide us with any number of ways of smoothing out differences to produce people who find happiness in conforming to a dominant culture. But should we want to impose, or even encourage, uniformity? Would it be better to foster ways of appreciating difference?

Plainly, the use of mapping and sequencing techniques to identify gene-environment-phenotype correlations in the fashion of traditional behavioral genetics will only take us a very small part of the way to understanding the causes of the human motivations and actions that most fascinate us. One large moral of my fables is that we can draw no firm conclusions about the role of the genes until we are able to fathom the mechanisms that link genes to behavior: Brute associations between input and output leave the most significant questions unaddressed. Now there is an alternative to the direct assault on complicated pieces of human behavior. Geneticists interested in dissecting behavior can begin more humbly. They can return to the fly and the worm.

Worms and flies do not immediately impress the uninitiated with their behavioral virtuosity. Nonetheless, they do some things that larger, more striking organisms embed in more complicated patterns of action: They coordinate sense and motion, they learn (albeit in a rudimentary way) from experience. Recent experimental studies of fruit flies have begun to reveal a fascinating picture of the fundamentals of long-term memory. Normal flies can learn to associate an odor with an unpleasant shock and will demonstrate their "knowledge" by fleeing the smell. By identifying a mutant allele in those "forgetful" flies who doggedly return for more, molecular geneticists have uncovered a cellular mechanism involved in memory, which, they suspect, is shared by other animals—like mice and very probably ourselves. Similarly, many researchers are excited by the prospect of coupling detailed knowledge of the sequence of the nematode worm, *Caenorhabditis elegans,* with the "neural wiring diagram" that already exists for this species, hoping that knowledge of patterns of gene activation in different neurons will illuminate basic mechanisms common to all creatures with a nervous system.

Groundbreaking work on memory in flies or locomotion in worms, however rigorous and imaginative it may be, will probably never compete for public attention with controversial speculations about genes for sexual orientation and aggression. 'Twas ever thus: Feeblemindedness was the hot genetic topic of the 1920s, and "thalassophilia" probably received more notice than eye color in fruit flies or the coat patterns of guinea pigs. If the history of genetics is any guide, then our best hopes of ultimately understanding the genetics of human behavior are likely to come from combining the careful study of simple systems with equally thorough and dispassionate research directed at human behavioral traits. Linking molecular genetics to developmental neurophysiology is not guaranteed to take us where we want to go; at some point the signposts from model organisms may be missing. That does not controvert the moral drawn above: Without attention to the neural mechanisms that cause behavior, we shall not understand what the associations between genotypes and phenotypes are telling us. Those who vault over the brain do not know where they land.

The trouble is that complex patterns of human behavior are connected with urgent social problems. Sex and violence do not lure scientists merely out of prurience. Politicians are prepared to pay for information that can be applied to help resist the centrifugal forces that split families or reduce the incidence of violent crime. Skeptical of ventures in social engineering, they want scientists to tell them how behavioral differences among people are traceable to differences among genes. No matter that the issues are complex, that firm conclusions are hard to come by. Given the importance of the intended applications, scientific niceties must give way to practical demands.

If the study of human behavior admitted two methods—one sure but slow, the other fast, promising, but potentially inaccurate— pressing problems might reasonably lead us to cut corners and take risks. That, however, is not our predicament. Refining classical behavioral genetics through the introduction of new molecular techniques might provide brute correlations between the presence of certain sequences and actions we hope to limit or to reduce, but we shall not know how to deal reasonably with people who carry such sequences until we understand how the behavior is caused. Overinterpretation of patterns of association is overwhelmingly likely, for the fallacies of genetalk exert a powerful grip. Like so many of our well-intentioned predecessors in the applications of genetics, we may very easily come to believe, incorrectly, that identifiable individuals are determined by their genes to engage in antisocial activities, and having no further idea of the causes of their behavior, we may feel that we have no option but to confine them, in advance, for the good of society. The end is worthy but the strategy misguided, for the proper protection of citizens requires understanding the causes of actions that threaten them and, for all its correlations, classical behavioral genetics does not deliver the causes.

Desperate times call for desperate remedies. Today there is a sense, strong in America, more muted but still present in other affluent societies, that mindless violence is rapidly increasing, that it must be controlled, that niceties of justice as well as scientific accuracy can be sacrificed to deliver people from the threat of random, unmotivated destruction. Urban parents grieve for children gunned down in the park, and their tears call us to prevent similar tragedies. Some think that genetic research is part of the answer; inspired by announcements of a "genetic basis" for violent behavior, they are happy to fund scientists to find the vicious alleles. Perhaps the ultimate idea is to develop a system of quarantine to protect others from genetically based "social sickness," compromising justice by segregating people before they commit crimes, compromising science by leaping to conclusions about genetic determinism, and very probably by misclassifying people into the bargain.

Exciting results will surely be forthcoming. Indeed they can be obtained without unpleasant fieldwork or bothersome experiments. For decades, geneticists have been intrigued by *pleiotropy,* which occurs when a locus exerts multiple effects on the phenotype. Pleiotropy is very common, occurring whenever a protein affects more than one phenotypic trait, as, for example, in genetic diseases with multiple symptoms. Those convinced that the roots of violence lie in the genes may want to finish composing the following possible contribution to the genetics of crime.

A hitherto unnoticed pleiotropy

The pleiotropic effects of genes at the violence locus, *Vil1,* have not previously been remarked. Studies by numerous authors and victims have revealed a strong correlation between the incidence of violent crimes in urban areas marked by decay. We therefore hypothesize that the mutant allele *Vil1⁻* not only causes its bearers to commit acts of violence, but also to seek out run-down city centers. It follows that *Vil1⁻* bearers have a tendency to congregate in regions of urban blight. Our preliminary data already confirm this consequence, although the mechanism for the attraction to decaying areas of cities is presently unknown . . .

Genetic determinists have to come to terms with the fact that crimes of violence occur disproportionately in grim environments, that the perpetrators and the victims are most often the inhabitants of such environments. The essay you are invited to complete offers the obvious solution: Violent criminals are driven to crime by their genes, and they live in decaying cities because the same genes draw them there. Most people regard the "solution" as absurd. Identifying the source of the absurdity teaches us something important about the errors of the genetic attack on crime.

The violence of slum life is a well-known phenomenon across cultures and across time, as real in Dickens's London as in the Johannesburg of yesterday or as in the South Bronx today. People who do not wear blinders confining their vision to genes and nothing but genes do not find the regularity particularly remarkable, taking it to signal obvious environmental causes of violent behavior. Individuals with no stake in society, with little to lose, aware of the contrast between their own dismal surroundings and the wealth that others enjoy, buffeted since childhood by hunger, neglect, and the anger of those around them, lash out at others, assaulting, raping, killing. Their circumstances are common to many, to the peaceful as well as the violent, to the victims as well as the assailants. What determines the difference? Perhaps more subtle environmental features—an impressive community leader here, a chance for a job there. Or perhaps there is indeed important genetic variation. It is not hard to specify the outlines of a plausible story.

An unknown array of causes probably sets up in each of us neural structures that determine our tolerance of environmental stress. Placid people, even when placed in desolate, threatening circumstances, will not be flooded with the hormones that trigger eruptions of violence. Others, with lower thresholds, might well be able to live serenely in the supportive surroundings that the middle classes enjoy but are moved to violence under the pressures of the inner city. Even if we assume that variations in the propensity to respond to environmental stress are entirely genetic, that a combination of alleles sets the threshold quite independently of environment, it is far from obvious that the appropriate way to attack crime is to identify the genetic differences.

An analogy will help to make the point. The manager of the rubbish dump has a problem. Fires periodically break out in the dump, and before they can be brought under control, they often spread to surrounding residential areas. One day, as he is pondering the difficulty, inspiration strikes. Not all discarded objects are equally flammable, and removing the most combustible would drastically reduce the risks of further fire. He commissions research to understand differences in flammability, carefully inspects the debris that is left at his dump, segregates and removes everything with a pronounced tendency to burn. The rate of dangerous fires is decreased, and the manager is happy.

But there was always a different solution, one that would have worked quite as effectively and would not have left him with the task of finding a home for the objects he removes. Fires break out because people are careless or, occasionally, deliberately destructive. Stricter control of the activities inside the dump and of access to it would eliminate the need for a division of objects according to their tendencies to burn. If the sources of fire are kept out, it doesn't matter how flammable the contents are. Provided the manager knows how to seal his dump and to ensure caution on the part of his workers, he can solve his problem.

By the same token, if we understood how to change the environmental pressures on those who experience the most bitter forms of urban poverty, there would be no need for a "genetic attack" on violent crime. Enthusiasts for genetic studies are likely to protest that we lack a precise environmental account of the causes of violent behavior. Indeed we do. They will also claim that genetics is a well-developed science, with established results, while sociology, by contrast . . . achieves less. Ironically, they overlook the counsel of pragmatism that is at the core of the hunt for "criminal genes": Our practical problems are sufficiently severe, sufficiently urgent, that we don't need an advanced scientific theory to solve them. Differences between the urban ghetto and the middle-class suburb are hardly subtle. *Commonsense* knowledge of the causes of violence and recognition of the stresses on all our lives and how they affect us tell us a great deal about why there is more violence in city areas that resemble bomb sites.

The stage is set for an important debate. Idealists believe that we can make considerable progress in combating crime by applying our rough, commonsense knowledge: Programs of job training, job opportunities, medical facilities, drug centers, and decent, affordable housing could transform the urban environment. Pragmatists counter that without greater understanding of the consequences of social measures there is little chance of solving the problem. Yet not even the most hard-boiled pragmatist should be attracted by modest proposals to identify the genetic differences among inhabitants of ghetto and slum, thus preventing likely aggressors from being burdens to themselves and their neighbors. Whether or not we know how to resolve the social problems, genetic substitutes will be no more efficacious than rearranging the deck chairs while the *Titanic* sinks.

The ideology of individualism is a powerful friend to genetic determinism. Genetic solutions reassure the successful: The rich man in his castle, the poor man at his gate (or, perhaps, in the dungeon), genes made them high or humble and governed their estate. Aspects of society that trouble us need not be traced to the breakdown of community, to the failure of affluent classes to find ways of alleviating the environmental desolation that callouses the spirit. There is ample motive for thinking that our temperament and our actions lie in our genes. And there is opportunity, too. Because of the ambiguities of genetalk, the compression of important distinctions in a convenient shorthand, it is easy to reach the seductive conclusions.

If twenty-first-century molecular genetics is to deliver its important results without plunging us again into familiar overinterpretations of the genetics of human behavior, we shall need to foster public vigilance. Perhaps it would be worth imitating the warnings printed on packets of American cigarettes; all popular discussions of genes must bear a disclaimer: "The International Union of Molecular Biologists has determined that casual thinking about genes is likely to cause misguided social policies."

Off-the-cuff genetalk permits us to localize responsibility in individuals—the antisocial are driven by their genes. Yet as we reflect, the picture changes. Whether we hold simplistically that human behavior is shaped by genes alone, or, more accurately, that it is the complex product of genes and environment, there are troubling questions about our freedom and responsibility. As utopian eugenicists probe the mechanisms, what becomes of the free moral agent? Where do we find ourselves?

12

Self-Dissection

Throughout the history of Western thought, the Delphic injunction "Know thyself!" has resounded with solemn significance, epitomizing for generations of inquirers the pinnacle of human intellectual concern. Probing our own nature appears enticing in distant prospect, more disturbing at a closer view. Freud was probably right in suggesting that science had delivered two major blows to our self-esteem: one when Copernicus displaced us from the center of the universe, another when Darwin revealed our kinship with nonhuman animals (although Freud may have overestimated his own impact in thinking that psychoanalysis had delivered a third blow of equal force). We no longer fear the vast empty spaces of the universe that appalled Pascal and his contemporaries, and only those who insist on a literal reading of scripture have failed to make their peace with Darwin. Our species, it seems, can come to terms with wounds to its *amour-propre*.

Molecular genetics offers no general revolutionary challenge to our self-conceptions as heliocentric astronomy and evolutionary biology did. Its threats are subtle and piecemeal, lying in the detailed articulation of a picture whose general shape has been clear—and disconcerting—for centuries. We face the possibility of dissecting the self, understanding the minutiae of the mechanisms out of which our most significant thoughts, emotions, and actions come. So long as the specifics are nebulous, the idea that human beings are no more than complicated physical systems is less challenging. When we confront an intricate causal explanation of how particular features of our lives are brought about through the interactions of molecules, will we be able to sustain our sense of the special value of our lives? Will molecular genetics, allied with developmental biology, neuroscience, and psychology, debase the currency of human life?

The fantasy about the causes of male sexual orientation that was introduced in the last chapter exposes the problem. How we are sexually attracted to others seems very important to our conception of ourselves and would be trivialized by the discovery that it has been shaped not only in the large, but in the fine texture of our sexual responses, by the particular concentrations of molecules present in particular regions of our brains at particular times in early childhood. At least Freud paid us the compliment of supposing that adult sexuality was the product of thoughts and feelings.

Ever since we came to understand diseases such as PKU we have known that molecular differences between people can have profound effects on their cognitive and emotional well-being. For want of a particular nucleotide sequence an enzyme is lost, the lack of that enzyme makes it impossible to metabolize phenylalanine, and the buildup of phenylalanine has devastating neurological consequences. However, this only shows that tragic losses can flow from tiny changes; it does not yet reveal that the variations among normally functioning people result from molecular distinctions, whether they stem from differences in the molecules delivered from the environment, from differences in DNA or in the proteins initially present in the fertilized egg, or from interactions among all these sources. So long as we can treat the molecular details as relevant only to the division between sickness and health, full human functioning and poignant losses, it is possible to preserve the view that we shape our own lives, that one person's joy in mountain vistas, another's delight in sailing or in the music of Beethoven, are not matters of molecular contingency but expressions of deep and important choices that we have made.

The details matter. We do not yet have them, of course, but perhaps in a century or so we may dissect an emotional response of special significance to some people. A personal case can stand for all—please change the example to center on whatever is important to you. Listening to Schubert songs or Verdi arias or, even more intensely, singing them, moves me with a sense of great emotional depth and completeness. The music captures me, leaving me with recollections of enormous spaces that I cannot describe in literal language. Imagine that I live long enough to hear the explanation of these experiences given by some future molecular biologists. There is a detailed account of the ways in which people come to associate certain sounds with patterns of neuronal activation, and it is shown how the particular pitches, timbres, harmonies, and rhythms used by Schubert and Verdi produce molecular effects in those, like me, who have undergone the right kinds of developmental histories. The effect is more pronounced if the auditory stimuli are reinforced through the use of lungs and larynx to produce the sounds; the appropriate neuronal patterns issue in the release of substances that are known to be associated with reports of deep emotion. As the researchers tell me, it is now completely understood how Schubert and Verdi produce their effects on those who have the right backgrounds. The great composers have, apparently, intuitively recognized how to answer Hamlet's challenge to Rosencrantz and Guildenstern: They can play upon me as on a pipe.

Where does this leave my sense of a deep emotional experience in hearing or performing Verdi and Schubert? After all, the biologists have shown me that it is simply a matter of the setting up of particular sensitivities that makes me release particular chemicals when I receive certain aural stimuli. Of course, I might have deceived myself all along, thinking that something significant, even sublime, was happening when I was simply exhibiting a complicated form of physical reflex. First-time parents know the puncturing experience of proudly displaying their baby's wonderful, rewarding smile and then learning from the more adept that it is simply the effect of indigestion. Perhaps my inability to articulate the significance of my encounters with Schubert and Verdi does not signal anything too deep for words, but just the difficulty of discovering meaning in a burp.

Skeptical doubts need to be tempered by two points, one concerning the complexity of the molecular processes that are likely to be uncovered in fathoming our own mental activities and one recognizing the possibility that molecular explanations need not displace our everyday psychological assessments. We have a tendency to think of a molecular account of the development of aural sensitivities (for example) as revealing that something we took to be extremely complex and mysterious is really very simple. Yet in the areas in which molecular explanations are already available, disclosure of the details fosters appreciation for the intricacy of life's strategies.

One of the most thoroughly understood developmental processes is the genetic "switching" that occurs in some viruses and, most prominently, in the bacteriophage lambda. (Bacteriophages are a kind of virus, receiving their name from the method of reproducing by destroying—"eating"—the bacteria they invade.) Lambda phages infect the intestinal bacterium, *Escherichia coli (E. coli),* and are able to multiply in two different ways. The first is a standard viral tactic: enter the cell, make numerous copies of the viral genetic material and package it in the appropriate protein coat, producing many new viruses; finally break open the cell and send the daughter viruses off to colonize new bacteria. Alternatively, and more insidiously, lambda phages can insert their DNA into the host chromosome, so that, as the bacterium divides, it carries with it the source of new viruses. To follow one strategy rather than the other, the phages must activate some genes and repress those needed for the alternative. Phages respond to environmental cues to "set the molecular switch" that determines their pattern of replication.

Although there are some gaps in contemporary understanding, molecular genetics has constructed a picture of how phages replicate, revealing an exquisitely timed sequence of interactions between DNA and proteins that involves more than a dozen independent major steps. One important stage involves a modification of the enzyme for transcribing DNA into RNA, in the presence of a particular protein, so that the enzyme effectively ignores a stop codon. Surrounding proteins also make crucial differences to the regulatory regions with which interactions occur and to the directions in which DNA is transcribed. If the molecular events underlying a basic developmental process in a virus with about ten genes are so intricate, why should we fear that the molecular explanation of our most significant experiences will shock us with a cheapening simplicity?

Discovering complex molecular processes underlying our thoughts and feelings need not displace our ordinary ways of conceiving ourselves. As the philosopher Daniel Dennett has lucidly argued, to explain a psychological state or process in biological terms is not necessarily to explain it away. After the molecular biologist has instructed me about the mechanism underlying my musical experiences, I might respond by reiterating my sense that they involve important emotions and insights. Would this simply be dogmatism, born of a childish resistance to admitting that I had deceived myself? Not if the molecular processes revealed the possibility of *psychological* accounts of my responses—either the ones I would always have given or others that would endow those responses with equal significance.

Naive parents are crestfallen when they learn that their baby's smile is a reflex response to the state of the intestines. Later, they will take pride in observing smiles that appear to be associated with awareness of their presence. The mere fact that some complex molecular process is involved in producing the smile has no impact unless it undercuts the idea that the smile results from the infant's delight in recognizing them. When the molecular biologists and neuroscientists explain the causal details, parents will be unperturbed so long as one of the complicated molecular states that brings about the baby's behavior can be identified as recognition of mother and father. Their initial pride in the first smile was dashed not because the explanation was physiological but because the physiological causes were quite different from the supposed psychological factor—joy in their appearance—that they had taken to lie behind the baby's beaming face.

Molecular accounts might leave in place the psychological explanations that give value to our experiences. The molecular biologists of my imaginary future point out to me that the development of the aural sensitivities that lead me to feel great elation or sorrow when I listen to Verdi involves some stupendous interactions among genes and proteins. Their accounts will not lead me to change my estimation of my musical experiences unless they are at odds with my preconceptions that certain patterns of sound have been associated with emotions of sadness, joy, and sympathy that I feel in my relations with other people, that my response to the music stems from an understanding of the human predicaments it conjures up, that the states which produce my emotional response are not only describable in molecular terms but can also be understood psychologically as thoughts and feelings about human situations. If the complexes that the molecular biologists describe are regularly associated with gastric distress, then I am lost—the experience was not what I took it to be. But if something similar occurs in me when I am thinking about various kinds of human experience or sympathizing with my friends, then my sense of the depth of the musical experience is vindicated. Far from subverting its import, the molecular explanation might disclose to me how various are the connections between the state that the music produces and other cognitive and emotive responses, showing the extraordinary way in which music condenses an array of human experience.

There are no guarantees. The molecular biology of the future might show us that things we take to be most important are brought about in mundane, even vulgar, ways. Yet probing the mechanisms behind our thoughts, feelings, and actions does not have to cheapen them. Perhaps it will turn out that the mechanism corresponds to our sense of the connection among our cognitive and emotional states—or, even if we are wrong, perhaps the connections are sufficiently rich to support a revised idea of the value of our experiences. Displaying the molecular underpinnings of our most precious thoughts and feelings need not reduce them to burps.

Yet does it matter, in the end, how our mental life is caused? Isn't the main point *that* it is caused—and that, in consequence, we cannot be said to act freely? Molecular biology thrusts into our faces a possibility that has alarmed reflective people since at least the seventeenth century. Struggle as we may to find significance in some of our experiences by showing that they are linked to other facets of our lives, we have to face the fact that all those experiences are caused, and caused ultimately by aspects of the world—the initial state of the zygote from which we develop, the environmental impacts upon us—over which we have no control. As we learn more about the molecular details of our lives, the question of whether we can retain psychological explanations that seem to lend value to some of our experiences should be only of secondary significance. The primary point is that what we call our "self" is fashioned out of quite external causes.

The impulse to pit human freedom against causation is very strong, both in the history of Western philosophy and in the thoughts of laypeople as they ponder the significance of their lives. We contrast the spontaneity we sometimes feel, the sense that our lives are, at least partly, our own, with the image of ourselves as puppets, jerked into movement in accordance with demands quite beyond our imaginings. The natural thought is that freedom is to be found where causation no longer reigns, but a few moments' reflection shows that this idea is much too simple.

Some natural events lack causes. When a radioactive atom decays, emitting energy, there is no further causal factor that brings about the event; the decay occurs randomly. Assemble a large number of radioactive atoms of the same type and you will find that some will decay very quickly, some will persist for a very long while, but most will decay within a period centered on a time peculiar to that element (the half-life). If you focus your attention on a particular atom and ask why it decayed when it did, physics can supply no answer. That is simply the way it chanced to be.

Maybe some events in our own lives take place when they do because of random occurrences on the atomic, or subatomic, level; indeed, this may happen in the beginning of a cancer that will ultimately prove fatal. The mere fact that parts of our existence are uncaused brings no comfort. If the molecular biologists conclude that some part of my development is a matter of chance, I shall not seize on this as the place in which my freedom expresses itself.

A dilemma threatens. If human actions are caused, they are not free. If they are uncaused, they are random and still not free. So there is no hope for human freedom. Eminent scientists have drawn precisely this moral, concluding that freedom is a delusion that the enlightened should repudiate.

Recognizing that evading causation does not help is a crucial step toward understanding how to construct a picture of ourselves within the view of the world offered by the sciences. Our apparent delusion is not that some actions are uncaused, but that they are caused in a special way. We envisage a ghostly part of ourselves that—somehow—interacts with the physical world, disposing our bodies to move in ways that correspond to its (or our?) intentions. This conception is decisively subverted by the news from molecular genetics, developmental biology, and neuroscience. For if the ethereal agent is to satisfy the demand that it is not the product of physical forces and processes, then we cannot suppose that it is revealed in the intricate unfolding that leads from the zygote to a conscious human being. Instead we must assume that, in some form or other, it attaches to our bodies very early (at conception?), interacting with the organic world in some utterly mysterious way. As we are able to locate physiological processes that are associated with complex mental capacities—and with losses of psychological function—the ghostly self comes to seem more and more like an invention designed only to soothe our wounded vanity, a hangover from more muddled ways of contemplating ourselves.

Yet perhaps we do not need an ethereal entity with questionable ties to the physical world, for what is surely important about the allegedly confused picture of human freedom is the fact that our goals, ideals, plans, and intentions are the sources of our actions. The processes of normal growth and development lead from a fertilized egg to an adult human being whose brain responds in unique ways to the impact of perceptual experience, sending signals to muscles that move the body to action. Sometimes the movements are forced, as when a person is pushed in a direction that she does not wish to take. On other occasions, the actions proceed from the agent's own wishes, as when the same person calmly and reflectively chooses an entertainment for the evening that suits her mood. Both types of action may be equally determined, admitting of detailed molecular explanations that show just how and why the neurons fire and the body moves. But the examples differ in that one, but not the other, involves states of the agent's brain that express what she wants and that play major causal roles in what she does.

One of the most incisive thinkers in the history of Western thought, David Hume, achieved a crucial insight when he recognized that what matters to human freedom is not *whether* an action is caused, but *how* it is caused, and that, to a first approximation, our actions are free when they are caused by our desires. Roughly, freedom is invaded when people are moved from without, when they are tugged like marionettes, and freedom is expressed when they are moved from within, when external pressures do not determine but are filtered through the states of the brain that express what the person wants. Suggestive though this conception may be, it cannot be the whole story. As we ponder the particular way in which contemporary molecular biology develops the basic scientific picture of ourselves, coercion threatens from within. Our adult wants, our plans and ideals, come to be as the result of processes that trace back to the state of the zygote out of which we grew and to the molecular character of innumerable environments. What consolation is it to think of our actions as stemming from the self, if the self itself is built up out of intricate interactions among molecules?

A sad but familiar fact about some actions is that they proceed from desires that hold the agent in their grip. People who struggle to resist their addiction to drugs may be overwhelmed by the craving for a fix, so driven that they do things they find repugnant, knowing that they will later feel intense regret; the teenage runaways who sell their bodies for crack are often, if not always, coerced from within. Simply acting from desire—catering to sexual tastes one finds disgusting, buying and taking the drugs—may be a matter of compulsion, not of freedom. Are all our wishes thrust upon us? Are they all compulsions? Or is there a difference between desires whose expression constitutes freedom and those that impel us from within?

Addicts who are trying to quit are genuinely torn. They have strong desires for the enticing substance and yet also deep wishes to resist. After the fix, as they consider their lives, they want not to be driven, they want not to want the drug. Many of them would like their lives to go completely differently; they aspire to be people who are untouched by the familiar craving and, liberated, able to satisfy other wishes—perhaps for a stable home, another community, opportunities to develop talents—with which the hunger for the drug conflicts. Two kinds of inclinations constantly do battle within them: one that accords with their sense of themselves and the overall plan they would like their lives to follow, the other an intense urge running counter to that plan. If the sharp, transient impulse moves them to action, we see them as compelled; if they are able to resist it, they seem on the way to regaining their freedom.

Suppose that you are forced at knifepoint to do something you find detestable. It would be quite wrong, not to say extraordinarily cruel, to suggest after the event that you wanted to do what you did. You preferred to perform the action than be knifed—and if the deed disgusts you enough, you may later feel ashamed, resenting your inclinations to conform—but the desires that are constitutive of the person you are, those that accord with your aspirations and conception of what would be a valuable life for you, are quite at odds with the action. Freedom is not a matter of acting from wishes that are thrust upon us, but of expressing those inclinations that are bound up with our identity.

Contemporary philosophers, building on Hume, have found ways of developing this perspective. Being free involves forming a conception of who one is. Many people, though by no means all, arrive at young adulthood with a sense of themselves and of the ways in which they plan to pursue happiness. Their free actions accord with desires that flow from this plan; they are limited and coerced by factors, either external or internal, that check its fulfillment. For many people, what is of primary significance is to be part of a family, embedded within a community, and their freedom is expressed in desires to serve others or to come together with others for celebration or commiseration. Probably a smaller number find their identity in their work or in their devotion to a cause, acting freely in ways that appear compulsive to those who think of work only as a necessary means. (The molecular biologists who spend all hours at their laboratories will often confess exhilaration at having something so significant to do with their lives.)

Do they deceive themselves? Convinced of the value of our own way of ordering our lives, we may look at others and see them as missing what is fundamental. Dedicated scientists might regard a life devoted to family or community as hollow and worthless, while those so devoted might view a relentless intellectual pursuit, with personal relations consigned to the margins, as a distortion of human existence. It is easy to take a further step, to explain the misguided self-conceptions in terms of the quirks of development. How sad that the homemaker never had serious educational opportunities! How sad that the clever child was cut off from a rich social life! Soon the familiar questions recur. However our conception of what is valuable to our lives develops, it does so as a result of contingent processes quite beyond our control. Innumerable molecular interactions shape the body and the brain, stitching together a person with a sense of self. If freedom is a matter of expressing a sense of self, and if, in turn, the sense of self is the product of an idiosyncratic history of minute physical happenings, how can we be free?

We should resist this line of thought. Grand rhetoric about human freedom seduces us into thinking that we must, quite literally, make ourselves. As we probe the details of our development, disappointment is inevitable, but the conception was always incoherent. If the self that allegedly makes itself is already fully formed, then it does not, after all, *make* itself; if it is not fully formed, then *it* does not make itself. To find our freedom, we have to start by acknowledging that we are the people we are because of events that are beyond our control, even beyond our understanding.

Some people do not act freely because they never develop, in any sphere of their lives, a sense of what matters for them. Others, while forming a conception of what is significant, seem to have it imposed upon them because of severe limitations in the lives they could expect to lead. Judgments that homemakers or research scientists have had their self-conceptions thrust upon them might recognize a lack of opportunities during development. Children sometimes grow up with parental expectations that they will take over the family farm or pursue a particular career, sometimes divide early into social groups that effectively eliminate certain choices. Certainly, a person whose early life is dominated by a culture that is indifferent to education and exalts the values of home and community seems less free than one who is faced with both developmental options and whose path reflects personal responses to them.

Human freedom is not, perhaps, an all-or-nothing affair. People vary greatly in their plans for their own life, in their senses of what is significant. Those who are luckiest grow up in ways that present to them a very large part of this range of self-conceptions, inviting them to choose what kind of life suits them. Their choices are not thrust on them but are expressions of who they are. It does not matter that the selections are caused; the important point is that the characteristics of the growing person face a rich array of options, and that those traits and the available possibilities jointly determine which course will be pursued. As the range of opportunities diminishes, the outcomes no longer reflect the fine grain of the person's character, until, in the limit, when there is just one available option, the sense of self becomes imposed.

Our freedom is not impugned by the fact that we are complicated biological systems, our adult states brought about through the modification of a zygote by a vast aggregate of internal and external causes (not all of which may be best understood by forsaking our everyday psychological descriptions in favor of the molecular complexities). We are free to the extent that the mind-boggling history of molecular events leads to the development of a conception of our own lives and what matters to them, that it involves the recognition that many alternative conceptions of significant lives are open to us, and that the intricate neural firings moving us to act constitute wishes and intentions that accord with our self-conceptions. Molecular biology promises to dissect the self, but we need not fear that it will remove from our vision those aspects that lend special meaning to our lives.

Pondering the more abstract "philosophical" questions of this chapter places earlier discussions of eugenic decisions in a new light. Terminating a pregnancy appears merciful when the child who would be born would inevitably experience extreme pain. In most of the cases of prenatal testing that have concerned us, however, the suffering of the child has not been central: Many of the most terrible neurodegenerative diseases allow for the control of pain with sedatives. If moral justification must appeal to human suffering, then, it appears, we can only point to the suffering of other family members. Is it permissible to abort a fetus to spare them anguish?

It is time to take stock. I have argued that the loss of genetic innocence imposes new responsibilities. To a limited extent we can already look into the future and foresee a few of the traits of our children. The explosion of molecular knowledge will make some form of eugenics inevitable. Utopian eugenics, the most attractive option, emphasizes that reproductive decisions should be taken freely by people with a clear view of the relevant moral parameters. My aim has been to defend utopian eugenics against important doubts and objections, to argue that abortion is a legitimate means of avoiding future human suffering, to show how we can resist the pressures to inflate the role that genes play in human traits, to indicate how we can retain what seems most valuable about our lives.

Yet the boundaries of the moral decisions utopian eugenicists commend have only been drawn impressionistically. I have suggested that the reflections that precede reproductive decisions should focus on the quality of the lives that would ensue, making intuitive though vague connections with the avoidance of human suffering. But the presence or absence of pain is only one dimension of the quality of a life. The tragedy of many genetic diseases is not that they issue in great pain—although some do—but that they prevent the development of a self. The possibility of human freedom is obliterated at the beginning.

Utopian eugenics seeks to prevent such tragedies, to promote lives that achieve genuine quality because they are the lives of free human beings. Appreciating the centrality of human freedom to evaluations of the quality of lives makes plain the compassionate role that molecular genetics may come to play in reproductive choices. With the ingredients assembled, it is now possible to develop a more systematic account of the considerations that should guide our descendants as they apply their newfound knowledge to make choices about nascent human lives.

13

The Quality of Lives

What makes lives go well or badly? All of us know (or know of) extreme examples. Our first thoughts naturally focus on happiness and achievement, on the one hand, pain and frustration, on the other, and we contrast those who live long, fulfilling lives (the great scientist who both wins the Nobel Prize and enjoys rewarding personal relationships) with those whose days are brief and agonizing (children who die young of diseases that palliatives cannot touch). But human lives are more than the sum of their pleasant or painful experiences. A life free of pain, one that even contains mild pleasure, may nevertheless be unfortunate: Children with neurodegenerative diseases, such as Hurler syndrome or Canavan's disease, may receive scrupulous care, preventing any disagreeable sensation; some people live their lives adrift, washed by a sequence of pleasurable experiences, never thinking about a direction, accomplishing nothing, and dying without regret. Conversely, sufferers from birth defects such as spina bifida may experience great pain but overcome substantial obstacles to create lives that bring true satisfaction.

As we move away from the most forceful examples, discomfort begins to set in; we start to doubt our ability to make judgments about the quality of others' lives. Were we too quick in evaluating the apparently aimless life that wanders from one sensual pleasure to another? The assessment that it "doesn't amount to anything" might all too easily embody a particular conception of human value, emphasizing achievement or deep relations with others, perhaps insisting that unexamined lives are not worth living. Fear of appealing to the quality of a person's life in making medical decisions, including decisions about abortion, surely rests on recognizing that rather specific conceptions of what is good for people quickly insinuate themselves into our thinking. Pronouncements about the quality of life might be dangerous because they guide our lives by lights that are quite contrary to our own. Generations of recipients have responded to the efforts of well-intentioned do-gooders by protesting that they "don't want to be done good to."

The history of human thought is dotted with once-influential doctrines about "the good life." Aristotle proposed that our lives go well to the extent that we are able to develop our natural talents. For the Church fathers, by contrast, our secular abilities were quite insignificant: The good life is one in which the human soul becomes prepared for union with God. Enlightenment thinkers changed the perspective, viewing the attainment of knowledge as the highest human goal and conceiving quality of life in terms of satisfaction for the inquiring mind. Reacting against the glorification of the intellect, the Romantics found human fulfillment in the development of deep emotions. Residues of all these varying ways of evaluating the quality of human lives are with us today, but using any of them, individually or in an eclectic mixture, would appear to ride roughshod over individuals' own conceptions of the values of their lives.

Contemporary democracies are historically rooted in seventeenth- and eighteenth-century campaigns for religious toleration. Fear of fanaticism runs deep in the cultures of today's affluent societies, and we are rightly skeptical that others can legitimately dictate to us the values by which we try to order our lives. Any adequate account of the quality of human lives must be pluralistic, acknowledging that people are authoritative with respect to their own lives and to the ways in which they find value. However, admitting pluralism does not mean that we must recognize all lives as being of equal quality. Different lives can go more or less well as judged by their own standards of success, and it is even sometimes possible to predict that a person's life will be of low quality whatever conception of value is adopted.

Fundamental to the approach sketched in previous chapters, and to be more fully articulated here, is the idea that a precondition for a life of even modest quality is an ability to form a sense of self and to formulate, for oneself, what matters. Instead of focusing on pleasure or the absence of pain, I take the core of a minimally valuable life to be a person's chosen ideal of that life's direction. Sometimes this is explicitly, almost self-consciously, articulated: The young hero gazes into the dim distance, squares his shoulders, and ventures forth to realize his ambitions. Far more frequently it is unconscious, implicit in a thousand decisions and actions, perhaps expressed retrospectively, with surprise and a sense of apology. So a woman who has dedicated herself for years to family and friends, caring and nurturing in innumerable ways, may one day unburden herself to an intimate, reflecting on her difference from her daughter who is making her way in the world and declaring, almost with embarrassment for such "highfalutin talk," that what has mattered and continues to matter to her is the well-being of her family. Achieving a sense of self and of the proper direction of one's life is not just for philosophers in their studies or for some privileged elite, but is common to all those who could sincerely respond to the question "What is your life about?" Sadly, there are some, such as my imagined hedonistic drifter, for whom the correct answer would be "Nothing," whose vapid lives receive no shape from actions, words, or implicit attitudes.

The sense of self can change during a lifetime, as we become aware of possibilities that had not figured in our early conception of ourselves. Like Saul on the road to Damascus, people sometimes repudiate their former ways of living, dedicating themselves to new ideals. In such instances there is no single standard according to which the success of the life can be judged—the converted believer regrets the fierce persecution of his youth. Often, the later values seem to take priority, and the earlier episode is seen as an unfortunate beginning for a life that eventually found its right direction, or even as a necessary preliminary stage in a spiritual odyssey. Most troubling are those lives that change their shape too late, leaving a person convinced that everything so far accomplished is futile, that life has been lived all wrong, but with too little time left to chart a different course.

Postponing until later how pleasure and pain, desires fulfilled and frustrated, relate to a person's conception (or conceptions held sequentially) of what matters to yield a measure of the quality of a life, we should concentrate first on the ability to form any such conception. Many of the judgments made in situations of prenatal testing reflect either the impossibility or the low probability that a person developing from the fetus would ever be able to form a sense of what his life was about. Tay-Sachs is a clear, extreme case. Doctors, parents, even many religious leaders agree on the permissibility of terminating pregnancies when the fetus is diagnosed as positive for Tay-Sachs, not because the baby will suffer pain—that can relatively easily be avoided—but because neurodegeneration will start before the distinctive life of an individual person can begin. Other types of cognitive and emotive deficits move us in similar ways, for they preclude the possibility of any personal conception of what is valuable and important. Severe examples of Fragile X syndrome, Lesch-Nyhan, neurofibromatosis, and other similar afflictions violate the precondition for a life of even modest quality. There is nothing to be done except to alleviate pain and discomfort, no aspirations we can expect to foster, no plans, however humble, to bring to fruition.

Even children who will die young may achieve an incipient sense of their own lives, not by explicit reflection on what is significant for them but through their attitudes toward those around them or their interest in particular activities. Others who live well into adulthood sometimes do little more than shape their lives according to the vague contours available to young children; adults with Down syndrome often appear to have taken charge of their lives only in the most rudimentary ways. For still other people, a genetic condition sadly limits their activities and thus the range of opportunities open to them. Those born with severe forms of muscular dystrophy cannot direct themselves toward goals that demand great exertion; some genetic conditions make it impossible for women to bear children; others, involving extreme malformations of the genitalia, preclude normal sexual relations.

Limitations do not always matter, for the options excluded may have no appeal. People committed to an intellectual quest can find the restrictions on their bodily movement secondary inconveniences, minor detractions from the main point of their lives. We envisage Stephen Hawking, voyaging through strange seas of thought, transcending his body's bondage. Yet even many of those who grow up knowing that certain kinds of lives are out of their reach may nonetheless chafe at the restriction, feeling that they have been blocked from the direction they would have chosen for themselves.

When we try to evaluate the quality of a human life, we should attend to three different dimensions. The first focuses on whether the person has developed any sense of what is significant and how the conception of what matters was formed. The second assesses the extent to which those desires that are central to the person's life plan are satisfied: Did the person achieve those things that mattered most? Finally, the third is concerned with the character of the person's experience, the balance of pleasure and pain. I shall examine each of these dimensions in order.

Considering future lives in prospect, we ask if the capacity for determining one's own values is precluded, and in the tragic cases in which genetic tests reveal no possibility for such self-determination, issues about termination of the pregnancy are as clear as they ever are. Next, it is important to ask about the maturity of the self-conception and the manner of its making. Will it be possible to attain more than the embryonic sense of self characteristic of young children, to achieve a sense of what matters, articulated explicitly, or implicit in the pattern of thoughtful deeds? Will that sense be formed through confronting a wide array of possibilities, so that what is chosen reflects an individual's particular character? Or will the options always be severely restricted, effectively imposing from without a conception of what is valuable?

Lives that form no conception of what matters have gone badly, lives with only an embryonic conception have gone less well than those that achieve something more developed, lives in which there is greater freedom in the process of formation go better. The character of a person's genes, product of accidents of meiosis or fertilization, can affect this dimension of the quality of life profoundly. The most feared genetic conditions, such as the early neurodegenerative diseases or the most severe disruptions of the genetic material, blot out the prospects for self-shaping entirely. Others (Down syndrome, Sanfilippo syndrome) limit the extent of self-development. Still others (myotonic dystrophy) restrict the array of opportunities, and thus the autonomy of self-formation. But as we try to make responsible judgments, we cannot look to the genes alone.

Freely forming a rich sense of what is significant hardly guarantees that things will go well. Lives that begin with high hopes often end in frustration and bitterness. The aspiring author turns away from mountains of pink slips to write advertising copy, the devoted spouse and parent is dragged through an acrimonious divorce and eventually ignored by resentful children, a lifelong attempt to hold together a fragile local community fails. Into each life, of course, a little rain must fall, but it matters very much where it falls. All of us can tolerate showers on the periphery. Many people begin their day with an eager look at the sports pages, hoping that their favorite team has proved successful—but even those dedicated to hopeless causes, passionately praying that the Red Sox will win the World Series or that Bulgaria will triumph in the World Cup, can (usually) cope with the inevitable disappointment. Storms at the center of our lives are very different.

The map that fixes center and periphery embodies our personal sense of what matters. Each of us has many wants, far too many to hope that they could all be satisfied, but they differ greatly in priority through their proximity to our conception of significance. For some people in affluent societies, work is a necessary evil, something that interferes with the important things, relations with friends and family and the enjoyment of leisure. Job-related successes and failures have no great import, except insofar as they affect the opportunities for improving family life or holiday activities. To be passed over for a promotion or to come to recognize that their contributions are not particularly remarkable would disturb them only to the extent that it threatened to remove the means to attain goals that they have set outside the workplace. Others, by contrast, frame their own happiness in achieving successes of quite specific kinds, perhaps completing a difficult task, or attaining a position of social distinction, or receiving large financial rewards. Their desire for a career advance or the public recognition of their accomplishments is central to their lives: Success will bring them elation, failure cannot be shrugged off with equanimity.

The quality of a life is measured, along the second dimension, by the degree to which central desires are satisfied. When a single self-conception dominates a person's entire life, there is usually little problem in identifying which desires are central, which peripheral. Lives marked by abrupt changes in direction can be harder to assess. On occasion, the later perspective dominates, the successes or failures of repudiated youth are irrelevant, and our evaluation turns on the person's ability to realize her more mature aspirations. Sometimes, though, it would be wrong to endorse a change of heart. Late in life, looking at the past through jaundiced eyes, someone may think of her earlier successes as worthless, dying with chagrin as she reflects on what she might have done. Sad as it is, the clouding of final hours may not tarnish the whole, and we may properly regard the life as one that has gone well, regretting only that a last depression prevented the woman from sharing that vision.

Genetic tests cannot foresee the shapes that future people will give to their lives, thereby constructing the map that locates their central desires, much less anticipate the contingencies that will affect whether those desires are fulfilled or shattered. Precisely this point underlies the complaint, already considered, that fascination with prenatal testing embodies an overprudential approach to life. Because the quality of a future human life will be affected by so much that we cannot foresee, why should we use molecular tools to determine whether or not a fetus carries some feared combination of alleles? We can now see that genetic tests are pertinent to uncovering potentially serious risks of two distinct kinds. The first, already examined, resides in the possibility that the life in prospect may be of low quality because of genetically based limitations on the ability to form a sense of what matters, or even that it may not allow for any process of self-determination at all. The second reflects the fact that those with only limited possibilities for shaping their lives may have predictably high chances that their central wishes will be frustrated.

Mobility and independence are important to most of us: Many of the central desires that flow from our conceptions of what matters in our lives could not be fulfilled if we forfeited our ability to control our muscles, which is why the loss of that control appears as a frightening possibility. To learn that a fetus carries a genetic condition that will lead to irreversible muscular degeneration in early childhood, that the person born would be able to move and to communicate only with the aid of equipment and considerable human support, would incline many prospective parents to conclude that the life would not be worth living. Yet is this not to impose values from without? People born with such disabilities develop self-conceptions whose corresponding central goals can be attained without independent mobility—they draw their horizons around their distinctive talents. Their accomplishments belie the old stereotype of patients with spina bifida or severe forms of muscular dystrophy doomed to lives of unacceptably low quality. But while being rightly impressed with what they have achieved, we should beware of concluding from prominent examples that there will always be distinctive talents to be developed, that excluding the entire array of personal goals for which independent mobility is required will leave something toward which a life can be successfully directed.

Some diseases of muscular degeneration are so severe as to permit little physical activity except when parts of the body are moved by other people. Perhaps the patient can speak, slowly and haltingly, with the aid of special equipment, but even basic activities of feeding and washing cannot be carried out independently. What remains around which a life can be built? Intellectual quests, of course, for the mind's motion may mock the inert body. The mind is, indeed, its own place, but only a very few find that its confines surround enough substance for a life.

Forewarned of unavoidable muscular decay, we recognize that the life in prospect must find its central concerns in intellectual pursuits and ask how likely it is that these will prove satisfying. The genetic information is pertinent not because it rules out the possibility of a life of high quality, but because it restricts the options of the person-to-be, allowing only for the development of central concerns that most people would be unable to satisfy. Most—but not all. Critics should be quick to note the existence of those, like Stephen Hawking, who live rich and rewarding lives while suffering just the types of physical limitations and dependencies I have described. So should we congratulate ourselves on formulating a framework for decision that would favor terminating future Hawkings?

Contemporary opponents of abortion often trot out the "Beethoven argument": the reproductive history of the Beethoven family, with its catalog of syphilis, stillbirth, blindness, and premature death, surely would have moved a thoroughly modern counselor to recommend curtailing the pregnancy, and so the world would have lost one of its most transcendent geniuses. The appropriate response is to point out that we do not know what great lives have been lost through the performance of other abortions—or because parents who carried a pregnancy to term did not abort and beget a different child. Provided that abortion is often followed by a new conception, there is no reason to think that the number of geniuses will be affected, unless there is some association between genius and the very risk factors that prompt decisions to abort. So far as we know, extraordinary abilities are quite independent of such risk factors and, in consequence, a policy to take risk factors seriously should not change the frequency of geniuses any more than it should alter the number of philatelists or of pathological liars.

Similarly, we have no reason to think that people who bear genes for degenerative muscular diseases are likely to have the intellectual gifts that enrich Hawking's life—or even that they are likely to have resources to sustain a life tightly centered on the intellectual. Just as we should resist the notion that our genes are our destiny, so too we should oppose the polar view: We are not infinitely plastic, capable of being molded in any direction by the appropriate environment. The thought that people born with severe degenerative muscular disorders that will confine them to bed or to a wheelchair from childhood on will—always? often?—be able to make satisfying lives for themselves by engaging with intellectual problems, so long as they are provided with a supportive environment, is as unjustified as comparably sweeping genetic determinist doctrines. We should be mindful of the woman who was paralyzed from the neck down and who petitioned the Quebec courts to allow her respirator to be turned off; after two and one half years of watching television and "looking at the walls," she protested that she had had enough.

When our most central wishes are fulfilled, we often experience joy and delight; when they are frustrated, we typically feel anguish and grief. But not always. A devoted family man might bask for years in the thought that his wife and children reciprocate his love, despite the fact that, with skill worthy of Molière's most ingenious social impresarios, his relatives and associates conceal from him the fact that he is widely regarded as a shallow, sentimental buffoon whose wife is routinely unfaithful and whose children are indifferent to him; even in death, he remains happily ignorant. Has his life gone well? In one sense, it seems that it has, for his experiences have been largely pleasant. Nevertheless, there are surely grounds for a negative judgment: All the satisfaction he takes in the appearance of domestic harmony is mocked by the reality; none of his central desires is actually fulfilled.

The perspective I have been developing deliberately downplays the character of a person's experience in assessing the quality of his life. Its verdict in the case of the hoodwinked husband is quite straightforward. The man's central desires have been frustrated, and so despite the bliss born of ignorance, his life has been an unhappy one. Perhaps this is too uncompromising, too severe, too Puritanical. Shouldn't an account of the quality of human lives make room for the character of our experiences, responding to the elation we sometimes feel, the pains we sometimes suffer, whether these experiences are what they seem or the stuff of fancy?

It should. Lives are not simply to be valued in terms of the number of units of rapture generated by the flux of experience, but neither can our pleasures and pains be ignored. The life of the hapless husband would surely have been worse if his illusions had been shattered, better if his fond images had accorded with reality. We do not have to place the objective and the subjective, the actual fate of our wishes and the felt experience, on a single scale. It is not necessary to decide if it would be better for desires to be satisfied, while we believed, miserably, that they had been unfulfilled, or for a world of fantasy to disguise from us the failure of our hopes. The character of subjective experience can be acknowledged as a separate, third dimension along which the quality of lives can be evaluated.

For a few people, like the imaginary husband or the well-meaning amateur scientists who periodically write to me, explaining that they have found the secret of the universe, there is an important disparity between feeling and accomplishment. Usually, however, separating our hedonic fortunes from the fulfillment of our desires would be artificial. Lives that go well in realizing their central dreams are typically filled with delight. The prospect of the goal blunts the discomforts endured in striving for it, and unless we have sadly misjudged, attaining it provides compensatory elation. Accordingly, earlier thoughts about the assessment of lives in prospect were one-sided, focusing soberly on self-development and the fulfillment of central desires but neglecting the likely flow of subjective experience.

Parents who have received bad news from a prenatal test frequently ask first about the probable sufferings of the child-to-be, hearing the reassuring response that pain can be controlled. Absence of agony is not enough. Despite the reassurances, it is natural to ask what kind of life it will be, meaning to inquire both about the possibilities for developing and satisfying central desires and about the joys that might come. What kinds of pleasures will be available to someone whose development will never advance beyond early childhood or who will be confined, very young, to a wheelchair or a bed?

The most devastating genetic disorders allow for very little, disrupting development so thoroughly that there is no understanding of the care given by others, no laughter at even the simplest joke. Other afflicted children can enjoy a more varied repertoire, listening to stories, delighting in television cartoons, and, especially, responding joyfully to those who treat them with love. A lifetime of simple pleasures may be enough—or there may be a time at which the child in the adult body finds the flood of hormones bewildering and oppressive, a time at which the immobile patient sickens of the diet of entertainment, barely extended since childhood. Sadly, the drugs needed to temper pain may themselves diminish the opportunities for accomplishment or for pleasure.

The diagnosis on the card records only a combination of alleles. Because of the vicissitudes of the relationships among genes, environment, and phenotypes, there is no graphic portrayal of the physical characteristics of the person who would result. Still less are the prospective parents offered a picture of how the person-to-be would determine her sense of her life, of how likely it is that her central wishes could be fulfilled, of the character of her subjective experience. They must guess at all that, guided by whatever statistical or anecdotal information doctors and counselors may provide. I believe that their reflections typically turn on an unarticulated assessment of the quality of the potential life, an evaluation that recognizes the dimensions I have distinguished—self-development, satisfaction of central desires, pleasurable experience. The consequences of that assessment will often be tentative, for the possibilities of misjudgment are obvious, and parents may wonder for the rest of their lives whether the agonizing decision was right. Correctness is too much to ask for. Despite the gravity of the consequences, they can only try to decide responsibly, and far from imposing values on others, serious reflections on the potential quality of the life in prospect are a major part of a responsible decision.

But only a part. Human lives cannot be considered as isolated atoms, sealed off from one another. If continued, the new life will have consequences. Most obviously, it will affect the parents themselves. Aware of the possibilities of selfish choice, they can easily come to feel that the impact on the quality of their own lives must be set aside, even that the focus must be exclusively on the potential for the person who would develop from the fetus. When there are direct implications for young children, whose needs would have to be subordinated, it is plain that this would be too myopic a vision. The fetus is not yet a person, not yet the bearer of rights that can conflict with the rights of elder siblings, already born. To accommodate the needs of the child who would develop from the fetus, the parents would have to risk reducing the quality of the lives of their other children, and those children's interests must surely be pondered as they make their decision.

Even when there are no previous children to consider, it would often be wrong for the parents to decide that they alone will be affected by the burden of providing the resources that will be required if their child is to have a life of even modest quality. Traditional eugenic arguments gain their bite precisely because effects cannot be confined so narrowly. Only a callous society denies to children born with severe disabilities the expensive support that is needed to make something out of their lives. To the extent that funds for social programs are tightly limited, support for such children diverts money that could be used to improve the quality of the lives of many others. One by one, the consequences of reproductive decisions are tiny: The birth of a single child with special needs barely registers on the scale of costs of medical and educational projects. Nor does overall expenditure on handicapped children and adults necessarily detract significantly from the quality of the lives of others—the projects that have to be sacrificed may produce only small effects or the society may be able to take on the extra expense. Nonetheless, because reproductive decisions do not occur in social vacua, responsible reflections must recognize that new lives impose demands: "If everybody acted as we are inclined to, what would be the effect on the well-being of other children and adults in our society?"

The agonizing choices balance different concerns. First comes the quality of the life in prospect, evaluated according to the various considerations distinguished above. Next are the implications for other lives, both directly and indirectly affected. Finally, there is the value of the fetal life already in process, a value that would be destroyed if the pregnancy were ended. "Balancing" should not suggest a precision quite alien to these decisions. People who have thought and rethought their predicament might long for a calculus, a method of assigning numerical values and combining them to reveal exactly which course of action should be pursued. But there is no scale of measurement, no algorithm to be applied. From a great distance, we might imagine that morally correct decisions would be those that maximized the total value of the outcome, perhaps subject to the constraint that the rights of sentient persons must not be overridden; those who came to the wrong conclusions would thus appear like students who have made calculational errors in some difficult mathematical exercise. Ambitious champions of utopian eugenics might hope to discover the precise principles for measuring value, and through programs of public instruction, bring about a world in which people usually arrived at morally correct decisions.

Modesty is better. If, as I have suggested, responsible people faced with the hard choice of continuing or terminating a pregnancy reflect on the quality of the life-to-be (its potential for self-determination, the chance that its central wishes will be fulfilled, its scope for pleasurable experience), consider the impact on other lives, and weigh the value that would be lost by ending the fetal life prematurely, that is all anyone can ask for. There will sometimes be clear cases on which, with the exception of those adamantly opposed to abortion, all would agree. When there is no chance of self-determination (as with the early-onset neurodegenerative disorders) or when a low quality of life for the person-to-be combines with large impact on the lives of others (as might occur when the fetus tests positive for a degenerative muscular disease and the parents are already struggling to do the best for existing children), we might expect that those who ponder the important considerations will reach consensus on abortion. At the opposite extreme, when the genetic test reveals only that the person-to-be is at high risk for a disease that will strike in adulthood (a greater than 50 percent chance of heart disease in middle age, for example), then provided that there are no other grounds for concern about the pregnancy, evaluation of the quality of the life in prospect is not likely to lead to a judgment in favor of termination. Concurrence of opinion on such cases might prompt us to conclude that the verdicts reached are morally correct—but I prefer to think that "correctness" consists simply in the fact that those who follow responsible procedures reach agreement.

Between the extremes lie the hard cases. Suppose that a prenatal test discloses the presence of a pair of mutant alleles for cystic fibrosis (of the common form, delta 508). Because of the variability of the expression of the disease, and also because of the possibility of unanticipated forms of treatment, the parents cannot be sure how restricted the life of the child (and the adult) would be. They can be confident that she will be able to form a sense of what is important to her life, and they know that providing her with the attention she needs will detract from the welfare of her elder siblings. But these are very loose limits within which to make an exceptionally consequential decision. If one couple declare themselves unwilling to prevent the birth of a child who would have chances for a rewarding life, while another, similarly placed, resolve with regret that they should not risk the well-being of their family, by what standard can we judge that one of the decisions must be wrong?

Perhaps in a seductive slogan: "If in doubt, choose life." To terminate a pregnancy is to end a life, and that might seem graver than any rival consequence. Poised between alternative courses of action, we might naturally reject the one whose consequences were graver. This line of reasoning attempts to introduce an asymmetry where none is to be found. The life ended is not yet sentient—no person is destroyed. Something of value will be lost, but that loss, while indeterminate, is comparable to the effects of bringing into the world a child who cannot even attain a life of modest quality. Too many parents have already heard the honest but chilling report, "We know that this genetic condition is not normal, but we don't know what its effects are," have decided that they must give the fetus a chance at life, and have watched, with regret and often remorse, a child eke out in a hospital bed a wretched existence of indefinite duration. Just as people disagree about hard cases, so too they diverge in their opinions about whether the end of fetal life is the most weighty of the consequences.

Confronted by the possibility of selecting lives, at least in a limited fashion, citizens of the affluent democracies have already begun to alter their attitudes toward human life. Human molecular genetics forces us beyond the piecemeal consideration of a small number of diseases, compelling us to clarify the sentiments that have guided people in their difficult choices. If my attempt to expose their implicit judgments is right, then there are serious consequences. We cannot just shut off concern for the quality of human lives in other problematic situations.

For almost two millennia, medicine has been directed toward curing disease and promoting health, and, accordingly, prolonging human life has seemed a medical imperative. Contemporary medical technologies for coping with the ends of lives, like the new techniques for assaying the genetic condition of a fetus, raise questions about whether we should always honor this command. Transplanted to a new world that was unimaginable when they were formulated, the ancient maxims need rethinking. The attitudes of many people are already at odds with them, but they continue to have at least a formal stature, even for those doctors who uneasily violate them.

If it is appropriate to determine the fate of a fetus by considering the quality of life of the person who would develop from it, then apparently we should also make decisions about the future of patients who are being artificially kept alive by pondering the quality of the futures that await them. There is, of course, a difference of the first importance. At the center of all the complex apparatus is not simply a human life, but a person, someone whose interests and rights must not be overridden. Sometimes the patient can express his own wishes in the matter, and if he indicates that he wants to live, then that is, of course, decisive. But what if he declares a desire to die?

The wish for death may be transient, the product of momentary depression or despair. It may also be a sober reflection on the prospects that confront him. Looking into the future, the patient sees a life of dismal quality, perhaps even something that mocks his previous accomplishments. His doctors may agree with his evaluation. Why must they resist his determination? Why strive officiously to keep alive?

Tacit judgments at the beginning of life are moved by the desire that the lives brought into being should have at least modest quality. How we die matters to us, for the ending of our lives can be radically at odds with our sense of everything we find significant. Many intellectuals, for whom the life of the mind has been primary, dread the prospect of their brain's decay and ardently wish for death before the disintegration of what they identify as themselves. Equally, those who have prided themselves on lifelong independence fear an ending in which they will be reduced to asking others for everything. If what is of principal importance to the quality of our lives is our way of structuring ourselves, of fixing what matters for us, then prolonging life can constitute a desecration far worse than death.

Desires for a "quick finish," "not to linger," "to die with dignity," are nothing new; they were expressed in my childhood by numerous elderly people, ordinary folk confronting their own deaths. New medical techniques have given these wishes a public face. The growing strength of the hospice movement, the tolerance in the Netherlands of doctor-assisted death for patients who request release from terminal disease, the unwillingness of juries (both in Britain and America) to convict doctors who have helped their patients die, the increasing interest in formulating instructions in advance so that it will be possible to understand and carry out our wishes, even if we are no longer able to express them—all these signal a widely shared sense that the mere extension of our lives is secondary to preservation of a minimal quality.

Blindly adhering to a code we have inherited from a very different age, contemporary medicine might continue to do whatever can be done to restore normal functioning. Each medical specialty might gallantly try to maintain its distinctive organ or system, whatever is occurring elsewhere in the patient's body. This would surely be a caricature of medical practice, a myopic application of sophisticated pieces of technology that has lost sight of the patient and the sense of her life. Approximations to the caricature already threaten us, inspiring the dying to flee the hospital for the hospice and the healthy to draw up living wills. Perhaps the time has come for us to acknowledge that we have already broken the code in our practices of prenatal testing and increasingly in piecemeal efforts to allow patients to die in ways that accord with the patterns of their lives.

Many people, perhaps most, have seen a friend or a relative linger for days, weeks, months, or years in a condition that the patient would have wanted fervently to avoid, with little or no expectation of any improvement. Some years ago, an old friend, a woman to whom I had been very close in my childhood, suffered a series of strokes that left her unable to speak, unable to communicate. While in England, I went with her daughter to the hospital in which lay the shell of the vigorous woman I had known. The daughter and I kept up a conversation with her in the hope, probably absurd, that she could hear and understand what we were saying and that she could receive some comfort from it. Perhaps there was a momentary tremor when I pressed her hand—I wanted there to be some gesture of recognition—but it is all too likely that I imagined it. More than two years elapsed between the incapacitating stroke and the lucky virus that brought the gift of death, two years in which her faithful daughter kept daily vigil, in which she, and the nurses, came to long for a chance illness that could be left to run a painless but fatal course.

Why was it necessary to wait? Just as prospective parents can recognize that the quality of a life-to-be would be so low that is better to terminate a pregnancy now, so too spouses, children, relatives, and friends can see that the patient lying before them is the residue of someone they loved, whose wrecked body belies the prior life's identity. Opponents of euthanasia will point out, quite correctly, that there may be abuses, that relatives with financial interests in someone's death or indifferent doctors might close a life too quickly, that their judgment might override the wishes that the patient, if conscious, would have expressed, and that their verdict allows no appeal. Plainly, trying to ascertain what the patient would have wanted is of first importance—and, in making the assessment, we must reflect on the ways in which the patient's attitudes may have altered in the course of recent suffering; plainly it is imperative to ensure that the decision to end the life is motivated by love and not by greed. However, we should resist the demand that the patient's wishes for the particular contingency must be explicitly expressed. Once again, we are in danger of being seduced by the suggestion that, where there is a doubt, however minute, we must choose life.

Nobody who had known my old friend would have had the slightest hesitation in expressing her wishes: She would greatly have preferred a merciful injection to days, let alone months, of living death. There was no piece of paper, duly attested, to confirm that, and even if there had been, it would have been possible to quibble that the circumstances she hoped to avoid were different from those in which her strokes left her. Euthanasia is indeed irreversible. But what exactly is the significance of that fact? Perhaps this: If she had died and we later discovered a document expressing her wish to go on fighting, no matter what, then we would have done a wrong that could not be put right; but if we had kept her alive, later discovering that she wished immediate release from a lingering condition, then we could correct our original mistake. This rationale is spurious, for it supposes that the harm done by postponing euthanasia, when that is what is wished for, is not comparable to the harm done through premature death, when the determination is to continue fighting. To keep patients alive may harm them just as irreversibly as to allow them to die.

While it is vital to try to act in accordance with what people wish for the continuation or conclusion of their lives, it is erroneous to suppose their interests are always best served by keeping them alive unless there is documentary evidence that they would have wanted to die. When there is no written articulation of their wishes, we are in danger of doing them wrong by erring in either direction, and we honor them most by striving to find a course of action that would accord with the sense of significance expressed in their past lives. Our judgments are fallible—just as those made by prospective parents may be mistaken about the quality of a future life. Holding human life sacrosanct is a more comfortable doctrine, one that sidesteps hard moral decisions with an all-purpose formula, freeing us from consequential choices to be made with imperfect information. For better or worse, we are beyond the simple world in which that doctrine was appropriate; the formula cannot do justice to the moral choices we are already making; the hard decisions must be faced.

• • •

An explicit shift to making judgments about the quality of lives central to medical practice and medical policy would have even more far-reaching effects. Within decades, if we are fortunate, citizens of affluent societies will be able to take steps to prevent the diseases that presently kill or incapacitate in middle age. To the extent that human molecular genetics proves successful, both in introducing tests that predict high risks for cancer and heart disease and in devising ways of acting to reduce the risks, reactive medicine, which attempts to control damage already done, can give way to preventive medicine. Yet in the end, whatever preventive steps people take, their bodies will eventually decay. If our descendants continue to demand that all possible efforts must be made to prolong their lives, then it is evident that the economic burdens of medical care will continue to grow as the costs of genetic testing and early intervention are added to the expenses of struggling against death postponed.

Abstract descriptions of the ways in which medical resources are dispensed, especially in America, easily induce belief that the times are out of joint. The great preponderance of the money spent on health care in the United States is used to prolong lives for very short periods (under about eighteen months), and a large proportion of this is spent during the last month of life. Faced with the bare numbers, most Americans agree that the distribution is unbalanced. Concrete policies aimed at lowering the costs of terminal care prove far more controversial. The British National Health Service's refusal to provide renal dialysis to patients over sixty-five with malfunctioning kidneys is frequently taken as a symptom of the ills of a centralized health care system.

Imagine that you were offered, once you had reached the age of reason, two options. One would abide by the traditional imperatives of medicine, so that you, along with everyone else, would be assured the most intensive efforts to extend your life: No matter how long you live or how expensive the costs of responding to the damage your body has suffered, the doctors will be there to employ the most advanced technological resources on your behalf. The alternative begins from the idea that medical resources will be deployed primarily to promote the overall quality of lives, to free people from limitations that would affect their determination of self-conceptions, to lower their chances of premature death or incapacity that would block them from fulfilling their central desires; prevention comes first, and after the costs of prevention have been met, residual resources are used to prolong those lives that can be expected to maintain their quality, to allow others painless deaths that fit, as well as possible, the structure of their lives. Armed with perfect knowledge of your genotype, and of the myriad assaults that the world will make on your body, you would be able to tailor your choice to your needs. But you are ignorant. You might be born lucky, a person who has no need of preventive medicine because the risks for major diseases are low and who could garner an extra year at the end of a long life by taking advantage of medical heroics. Or you might be less fortunate, someone whose genotype and environment place at high risk for debilitating disease and whose range of options would be greatly expanded or whose healthy years significantly extended by means of preventive strategies. Unable to tell, you must play by the odds.

Even without detailed calculations, even without foreknowledge of the state of molecular medicine decades hence, it is not hard to decide. Current returns on the huge investment of resources late in human lives yield so little of high quality that we could probably retain the uncontroversially beneficial treatments while adopting a thorough preventive program. Or, less impressionistically, the gains from the first, traditional, option are slight if you are fortunate; the benefits from the alternative are considerable if you are unfortunate; since there is an appreciable chance that you will be unlucky, you do better to choose the second. Far from being the epitome of the faults of "socialized medicine," the British approach to renal dialysis is a small step in the direction of a rational medical policy grounded in proper concern for the quality of human lives. Medicine should be reoriented toward expanding the potential of the young. It should aim to provide us fourscore years of vigorous life, in a pattern of our own choosing—and a decent exit.

Human lives are no longer simply given to us, so that we must fight rearguard actions against their vicissitudes. Partial and imperfect though our knowledge is, we already have the power to shape lives, and in committing our resources to help those who cannot yet express their own wishes, we are bound to make judgments about the quality of lives. Our predicament is quite different from that of the ancient Greek doctors who wisely formulated goals and maxims appropriate to their age. Their code has been outgrown by our technological successes. Contemporary medicine must rest on different foundations, falling into place as an instrument for promoting the quality of human lives. But it is by no means the only such possible instrument.

14

An Unequal Inheritance

This book began with the patients of Children's Convalescent Hospital in San Diego. Those children, especially the boys and girls with the most severe genetic diseases, symbolize one important, morally praiseworthy motivation for applying molecular genetics: Lives, we feel, should not go like this, and if we can foresee them and forestall them, then it is right to do so. Earlier chapters have traced this intuitive feeling to its roots, to a concern for the quality of human lives as it is expressed in decisions taken by individuals to close off lives that must inevitably be of low quality, before there is a sentient person. Biomedical technology helps to make those decisions possible and, quite reasonably, many people are prepared to support its development and to pay for its merciful applications. As we have seen, any such applications must be accompanied by a public determination to support and treat with respect those afflicted with the genetic conditions we attempt to predict and to prevent.

Less than ten miles from Children's is another institution, Loma Portal Mental Health, a public facility accommodating a small number of children with behavioral, cognitive, and emotional problems. Some of the boys and girls have genetic conditions whose effects are amplified by the environments in which they have been reared; others are genetically indistinguishable from their peers who throng school playgrounds around Southern California. For a few weeks, or maybe months, the staff at Loma Portal will work hard with them, diagnosing trouble, prescribing medication, offering support and stability. Restored to normal functioning, the children will return to the situations from which they came, to broken families, often to a world of drugs, violence, and prostitution. A doctor who works with the younger ones acknowledges that the prospects for long-term recovery are very poor: Perhaps pessimistically, she sets the success rate at 0 percent. Although Loma Portal provides an interlude of care, the pressures of callous and brutalizing environments will eventually take their toll, the psychic wounds will deepen, and the children will slide beyond reach.

The contrast between these two institutions and the tragedies they represent provokes a question that fuels the fiercest denunciations of applications of molecular biology in general, and of the Human Genome Project in particular. Concerned with the quality of nascent lives, the United States and other affluent democracies are prepared to commit public resources to fighting the devastating diseases that afflict a small number of children. At the same time, critics charge, there is no serious public dedication to programs that would alleviate the plight of many, many more whose prospects are blighted by social and environmental causes. What accounts for the difference? If our fundamental goal is to promote the quality of human lives, does it matter that the potential is lost because of a malfunctioning enzyme, because of negligent and abusive parents, or through the other hostile forces of the ghetto and the slum?

These questions have been an important part of public debate. Enthusiastic rhetoric about the marvels of tomorrow's medicine, which makes it appear that tackling genetic disease is *the* important problem, invites critics to attribute uncharitable answers. Do the champions of molecular medicine fixate on the genes because they believe that genes are destiny?

Without embracing genetic determinism, it is possible to recognize that mutant alleles sometimes play a role in limiting human lives. Defense of the Genome Project does not rest on the claim that it is a universal panacea: Defenders can concede the need for caution in attributing power to the genes while emphasizing the promise of local good.

Or perhaps the critics charge that a genetic focus represents self-ishness and insensitivity. Attending to the genes and their malfunc-tioning can be viewed as a cheap way for well-off people in affluent societies to guard against the possibility that their children will be doomed to miserable lives. Those of us who live comfortably worry less that our sons and daughters will suffer from neglect or abuse—we, after all, intend to provide them with safe and nurturing environ-ments. But we are not immune to disaster. The genes may strike, and if they do, all our efforts will be in vain. Accordingly, we are very in-terested in one kind of cause of the diminution in quality of people's lives. Other kinds of causes, the environmental factors that wreak havoc with lives, are not (perceived as) our problem.

The discussions in earlier chapters have emphasized the genuine problems that must be overcome if the benefits of molecular genet-ics are to be spread across all social classes, but they have also indi-cated ways of solving those problems, at least in principle. The genetic focus is not necessarily the result of a one-eyed compassion, as if the only concern were to protect the quality of the lives of upper-middle-class children. There is a better answer to the critics' opening question, the answer of pragmatism. Moved by the predicaments of many children, whose lives are blighted in different ways, it is en-tirely reasonable to welcome a program offering realistic chances of doing good while resisting suggestions to invest in other ventures which have little prospect of success.

A pragmatic defense rests on an assessment of the future of mole-cular genetics and its applications whose nuances should be clearly understood. At present, because of difficulties we have canvassed ear-lier, the ability to predict phenotypes from genotypes is limited. For all the imprecision of current understanding of the social causes of psychopathology, clinicians can often foresee with great certainty that returning a child to a particular "home" environment will lead to a relapse within a few months. Thus, predictive exactness is not the source of the pragmatist's conviction that genetic causes are more tractable than their environmental counterparts.

Nor does the difference lie in our *present* ability to reverse the effects of genetic misfortune and a corresponding helplessness in the face of environmental causes. Molecular genetics currently offers the chance of preventing lives whose quality would inevitably be low and of repairing damage in a small range of cases: The rest is promise. Nevertheless, many reflective people, recalling the extraordinary successes of molecular biology, and comparing them with the recent history of attempts at social policy, believe that the promise will be actualized, that we shall be able to combat the disruptions of defective proteins, but that, lacking comparable expertise with the complexities of social causation, we are doomed to continue blundering (at least for the foreseeable future) from one misguided policy to another. Pragmatists need not lack sympathy; they may be guided instead by a sober appreciation of the limitations of our knowledge.

Our lives are the products of many lotteries, and only one of them shuffles and distributes pieces of DNA. Behind the often acrimonious controversy about the value of molecular genetics is a deep disagreement about the implications of this fact, a disagreement dividing pragmatists from idealists. Their dispute intertwines two large classes of questions. What is the extent of our obligation to aid people whose initial circumstances greatly reduce the quality of the lives they can expect to lead? What are the practical possibilities for meeting this obligation, specifically for fighting the environmental causes of pinched and painful lives?

These questions are vast, involving numerous complicated issues about the social causes of inequality as well as central problems in moral and political theory, and any serious attempt to answer them would require (at least) another book. Yet they cannot simply be ignored, for how we approach them affects the proposals and arguments of earlier chapters. I have been trying to show how the challenges raised by our increased understanding of human genetics can be addressed: Are these recommendations simply abstract possibilities that are quite at odds with the social and political realities of our world?

Idealists are moved by an analogy that starts by juxtaposing two things: First, the fact that our lives are in large measure the products of lotteries, both genetic and environmental, which deal fortunes unequally; and, second, an attractive social ideal, the ideal that people should have equal opportunities to live happy and rewarding lives. Mismatch between ideal and reality prompts citizens of affluent democracies to believe that justice demands some attempt to remedy the unequal accidents of birth and childhood. Some type of help is required. Help typically calls for money, and that requires some form of redistribution of resources. If we are prepared to reallocate assets to combat the effects of disabling alleles, then, by analogy, we should also be ready to undertake similar schemes to prevent the damage done by brutalizing environments.

Even those most devoted to the principle of individual autonomy, those who resist attempts to compel people to give up assets for the benefit of others, find it hard to oppose all schemes of redistribution. Children who are born with severe afflictions require compassionate care if the quality of their lives is not to be excruciatingly low, and only those in the grip of a dogmatic ideology can grudge the costs of providing for them. Respect for individual liberty is a worthy ideal, but that ideal cannot properly be expressed in a "hands off" attitude toward redistribution of assets. For if, as I suggested in an earlier chapter, our freedom is a matter of degree that depends on the extent to which we have the opportunity to be guided by alternative possibilities, then redistribution might decrease the autonomy of the privileged by only a slight amount while greatly enhancing the autonomy of the hopelessly disadvantaged. Far from manifesting a concern for the autonomy of all, the directive not to demand assets from the well-to-do would be more accurately advertised as a maxim to respect the liberty of the winners in the lotteries that fix initial circumstances.

But how far should we go? It is surely right to ask affluent, healthy people for funds that will provide compassionate care for children born with the most terrible genetic disabilities. Even if the quality of the lives of those who already live well could be improved by allocating the money elsewhere—say by building wonderful public sports facilities—any proposal for stripped-down "compassionate" care would be morally obscene. In earlier chapters I have argued for public subsidies to support those whose lives could be transformed by expensive forms of medical intervention, like the PKU diet and the many other weapons for combating genetic disease that we hope will result from biomedical research.

Some of the children whose lives have already been devastated by the environments into which they have been cast need, and receive, the same levels of compassionate care as those who are victims of their DNA (recall the boy whose brain was destroyed by battering). Yet idealists also believe that there are many others who, as things now stand, are likely to have lives of very low quality. The children of Loma Portal represent a vast group whose prospects could be transformed if their bleak and brutalizing environments were changed. If we feel the obligation to help the PKU children and others threatened by abnormal alleles, then it is surely because we cannot in good conscience stand idly by while human potential goes to waste. By the same token, are we not obliged to try to undo some of the terrible effects of the environmental lottery? It is tragically clear that the number of boys and girls doomed to lives of very low quality by the savage environments into which they are cast is much larger than the number afflicted with genetic disease or disability—and it is even possible that more die in childhood because of the hostile forces of the ghetto than from the results of defective alleles.

The pragmatists' response intertwines a number of considerations. Most prominent is a pessimistic assessment of the effects of social intervention. Reflecting on previous attempts to care for the disadvantaged, many people conclude that the problems are intractable, that well-meaning efforts to engineer solutions often fail to help those who are supposed to benefit, and, worse, have unforeseen consequences that generate further social disasters. The road to social catastrophe may be paved with the best of intentions. Additionally, pragmatists may stress the inefficiency of typical social interventions, claiming that, even where policies do bring benefits, they require expenditures much larger than the gains achieved. This thought alone would not prove telling—for, after all, we are willing to provide expensive care for victims of the most devastating genetic diseases, despite the fact that their quality of life is raised by only a small amount. However, where inefficiency on a small scale may be tolerable, inefficiency on a far larger scale might prove ruinous. Precisely because there are far more children who suffer because of desolate or threatening environments than children with the genes for PKU, or even all the children who are victims of genetic disease, we cannot take steps to alleviate their plight without greatly debasing the lives of the more fortunate members of the population.

Idealists have a number of ways of trying to counter pragmatist pessimism. They can disagree about the consequences of social interventions. Further, disclaiming the need for any sophisticated social science, idealists may reasonably point out that we know the sorts of things children need—better schools, safer streets and playgrounds, stable homes, and realistic prospects for jobs. Infusing money into job-training programs, teachers' salaries, school renovation, group homes for neglected and abused children, and developing an effective police force that is trusted by local residents would make an enormous difference to the expected quality of many lives. Is it really plausible to suppose that *nothing* could be done without greatly lowering average levels of well-being across the society? (Pessimism about the likely impact of social programs is, of course, comforting for those from whom sacrifices might otherwise be required.) Finally, in response to the charge that there may be unanticipated harmful consequences, idealists may suggest that grave problems require us to take risks, just as we might consider risky interventions when confronted with a devastating disease.

Prior to the wrangling over the likely consequences of social programs is a moral question. Reflection on the genetic cases teaches us that small losses in expected quality of life for healthy, well-to-do people are legitimately required when we can ameliorate the sufferings of the least fortunate and transform lives that would otherwise have been stunted so that they flourish almost as well as those of the healthy. A whole spectrum of moral positions is consistent with this judgment. At one pole are people who emphasize the fact that *only small sacrifices* are required to improve the lives of the genetically unlucky. At the other are people who stress the possibility of *improving the lives of the unlucky* and are glad that the sacrifices, in the genetic case, are only small. Between the extremes are all the ways of balancing concern to provide opportunities for all with the determination to maintain the average quality of life at a high level. There is a *moral spectrum* along which different people, including pragmatists and idealists of various stripes, place themselves differently.

Molecular genetics does not confront us with the difficulty of positioning ourselves on this moral spectrum. When we turn to evaluate the idealist analogy, however, disagreements about the pertinent social facts and about the proper position on the moral spectrum are interwoven. Many idealists would surely offer an explanation for the apparent recalcitrance of social problems. Our failure, they believe, can be traced to the limitations of a moral perspective that embraces the redistribution of assets only timidly, withdrawing when our efforts encounter obstacles. Perhaps the remedy for our ignorance is to try a variety of approaches and learn which ones are beneficial. That can only be done, of course, if we are willing to commit substantial resources, if we are prepared to position ourselves further toward the egalitarian end of the moral spectrum (the pole that emphasizes the provision of opportunities for the unlucky).

I have been exploring the most straightforward way of developing the idealist analogy, but the most straightforward way is not necessarily the right way, and the comparison so far considered may make it easier for pragmatists to dismiss idealists as starry-eyed optimists. In presenting the genetic example, I suggested that we are morally committed to supporting children whose quality of life could be significantly improved by applying our molecular knowledge; for example, children carrying the alleles for PKU. However, PKU is a parade case; the goal of molecular medicine is to develop similar ways of blocking the threats posed by other mutant alleles (and, we may hope, the coming decades will bring not only exercises in imitation, but also a commitment to provide the social background needed to make the molecular miracles work). The promise of applications of molecular genetics is real, but it is likely to issue not in sweeping solutions to problems of major families of disease—total victory in the war against cancer, say—but in a motley of techniques, useful to varying degrees in a broad range of cases. The idealist analogy can be developed differently by combining the concern for the quality of the lives of the least fortunate with a counterpart for this therapeutic pragmatism.

Statistics on rates of violent crime, unemployment, drug abuse, and low education levels in some regions of urban America are sufficiently horrifying to inspire idealists to call for sweeping plans of social action—or, perhaps, sufficiently numbing to cause pragmatists to resign themselves to the impossibility of tackling problems of such vast scope. Yet it is worth taking a look behind the statistics. In a moving book on the lives of some young boys in the Chicago housing projects, a book aptly titled *There Are No Children Here,* Alex Kotlowitz exposed quite specific needs that could be addressed very directly by a commitment of public funds. The public defender who represents one boy does not have time to listen to his story because of the overload of cases; the energetic and inspiring teacher who encourages the boy's younger brother has inadequate funds for books and supplies; parts of the projects are always dark because the city has given up trying to devise a system of lighting that would be van-

dal-proof; and, perhaps most importantly, the resourceful adminis-
trator who successfully sweeps a few buildings for drugs and guns
(earning the respect and gratitude of the inhabitants) is unable to extend his operation to twenty other complexes, because "money from
the Department of Housing and Urban Development has not been
forthcoming." It is hard to maintain that these difficulties are any
more insuperable than those biologists face in trying to repair the
damage done by defective, or missing, enzymes—and hard not to
believe that solutions to these mundane problems would bring real
improvements in the quality of human lives.

Although idealists and pragmatists may find some common
ground in the thought of redistributing resources for parallel programs that seek specific local ways of reversing the depressing effects both genes and environments have on the quality of lives,
significant moral differences are likely to linger. Because of the scale
of the social problems, there will be disagreements about how far societies are obliged to go, about how to balance concern for those who
are worst off with avoidance of society-wide losses in quality of life.
Indeed, idealists and pragmatists may divide among themselves about
the prospects for piecemeal change. Some members of both camps
maintain that the documented plight of children born into the most
brutal environments in the "advanced" world reveals that nothing can
be done by small increments. Pragmatists draw the conclusion that
nothing can be done. Radical idealists campaign for large social reforms. While they argue, the lights in the Chicago projects continue
to go out.

The debate between idealists and pragmatists is multifaceted—
perhaps more many-sided than is commonly supposed—and I do not
claim to have done more than scratch its surface here. But I believe
the idealist analogy cannot simply be dismissed. If the analogy does
nothing else, it should force each of us, privileged citizens of affluent societies, to reflect on the extent of our obligations to promote
the quality of lives that are currently bleak, to think about our own
chosen place on the moral spectrum.

The idealist critique reminds us that the problems with which this book has been concerned arise in a wider context. Champions of molecular medicine commonly believe they can achieve its benefits without attending to the broader social inequalities. It is easy to see the implications of the new human genetics as local: We can aspire to treat or prevent serious diseases and disabilities, and we must take quite specific steps to ensure that we do as much good and as little harm as possible. So, as early chapters suggested, we shall need to be cautious in introducing genetic tests, requiring that detailed statistical information should be available, instituting effective counseling, limiting the impulse to test on a grand scale without considering the impact on the lives of those who are tested. To safeguard the rights of people whose unfortunate alleles might make them outcasts, societies should restrict the uses that employers and insurers can make of genetic information. Indeed, as I argued, the vulnerability of a significant minority in future generations (the roughly 10 percent with genes that make them "uninsurable") underscores the need for a system of health care available to all. The opportunities to use DNA analysis in the fight against crime must be shaped by a clear understanding of the dangers of wrongful conviction, ideally through the construction of national databases. In all these instances, there are real dangers, but also definite procedures for avoiding or minimizing harm.

Yet from the earliest chapters in which these questions were discussed, it was obvious that there are plenty of loose ends. Will existing forms of discrimination—economic or racial—make it impossible for some people to benefit from the new biomedical technologies? Will the disadvantaged be harmed because solutions that are available in principle are not put into practice? Will we use the new biology as a forensic tool without threatening the residual rights of those who are already vulnerable? Many thoughtful critics are concerned that piecemeal fixes for the problems of applying the new knowledge justly and humanely will prove inadequate in societies resigned to accepting broader social inequalities. There is no necessary connection between resisting the idealists' call for a more sweeping redistributive program and opposing the measures I have recommended on the grounds that they are too costly. But we ought to wonder if societies whose policies reflect a position on the moral spectrum at a far remove from the egalitarian pole will decide that they can manage perfectly well without the expense.

These issues become more urgent in light of the second half of this book. I have defended the moral legitimacy of the enterprise of choosing our descendants. Once we lose our genetic innocence, we are committed to some form of eugenics. Emphasizing the importance of reproductive freedom, I have tried to show how many serious concerns about a future eugenic program can be addressed. It is morally justifiable to use abortion to prevent lives that would have been doomed to misery; indeed I believe that decisions to terminate a pregnancy sometimes show the greatest respect for human life. Nor should we think of our descendants as necessarily swept along by social prejudices, eradicating whole classes of people. They can avoid the determinist trap of overinterpreting the role of the genes, they can maintain a sense of human freedom and of the special value of human lives and experiences. Their decisions, I have suggested, should result from consideration of the quality of lives, both those that have already begun and those that might come to be, and in making the appropriate assessments, they should consider the possibilities for freely choosing life plans, the chance of satisfying the desires central to those plans, as well as the balance between pleasant and painful experiences.

In response to understandable concerns about the enterprise of choosing people, I have elaborated a program of utopian eugenics, which seems the most attractive option for implementing our knowledge of human genetics. But is it an option for a fairy-tale world? How will Modern Civics actually be taught in 2070? In the United States today and increasingly in the larger nations of Western Europe, people in certain segments of the population have virtually no chances for employment—or if jobs are open to them, they are menial and low-paying. To talk of the quality of life in the language of earlier chapters, stressing "self-determination" and "conceptions of what matters," should seem to them like a bitter joke. Deprived of any legal opportunities for pursuing lives that could plausibly seem worthwhile, their only sense of themselves can be one of rebellion. The language of utopian eugenics, with its refined thoughts about reproductive responsibility, is utterly remote from their struggles.

Prevailing social conditions also provide fuel for familiar eugenic arguments, perhaps presaging some future slide from utopian eugenics to the unwholesome practices of the eugenic past. Aware of the brutal realities, of the fact that children are born in threateningly large numbers to mothers who are hardly beyond puberty and whose scant receipts of love do not prepare them to give, it is easy to question the utopian ideal. Appalled by the numbers of children born into want and worse, even thoughtful people feel the force of an old eugenic complaint: High birthrates among the unproductive threaten the social and economic health of the nation. So we might declare, in the spirit of Justice Holmes, that three generations of welfare mothers are enough, that it is time to put an end to public support of the feckless.

Contemplating the problems of the urban poor, politically sophisticated commentators have become convinced that the difficulties are intensified by the numbers. Tracing high birthrates to programs that provide incentives to motherhood, some clamor for a reversal, proposing to cut off aid for women who have too many children. They may draw on an analogy, one very different from that which moves idealists. Isn't the foreseeable solution to problems of genetic disease to identify afflicted fetuses in advance and to prevent the births? In the same fashion, wouldn't withdrawal of aid from the overprocreative be a means to prevent lives whose quality would be low? Earlier discussions might even inspire an extension of the analogy. Attempts to decrease the number of children whose circumstances are desolate could be combined with respect for those actually born poor. Emulating the Cypriot treatment of thalassemia, future policy makers might find that the lives of the small number of children born to the indigent could be dramatically improved.

Nobody should be seduced by the suggested analogy—there are crucial differences between the two cases. Preventing births through abortion does not harm the person whose life is terminated, for there is no such person. By contrast, a social policy of withdrawing aid from mothers who "breed" too freely would surely injure innocent victims; the children they bear would experience even more extreme destitution. Indeed, the social policy would be analogous to prenatal testing with selective abortion only if there were a *person* whom attempts at abortion were to harm—that is, if the procedures for terminating pregnancies often failed but caused additional damage and mutilation to the individuals who survived. Equally fundamental is the difference between providing prospective parents with the opportunity to make informed decisions about whether to continue a pregnancy and coercing the indigent into abstaining from further reproduction. Throughout earlier discussions of eugenics, the principle of reproductive freedom has served as an important bulwark, dividing an apparently benign future from the excesses of the past. Withdrawing support from poor women is not as blatant a form of coercion as the method used on Carrie Buck, but it is nonetheless a breach of reproductive freedom.

As the political commentators rightly perceive, the numbers count, sometimes inspiring us to feel that the situation is so desperate as to demand policies that would ride roughshod over individuals. From a distance, it is easy to deplore the reproductive irresponsibility that deepens the problems of the urban poor. At closer range, it is much harder not to empathize with a woman's desire for small snatches of affection in a world that offers her nothing else, no way of shaping her life outside of procreation. Shock tactics that threaten such women with even deeper poverty if they continue to "breed" both deepen the sufferings of their children and deprive them of the last vestiges of their freedom.

Utopian eugenics envisages a world in which reproductive decisions are made freely *and responsibly*. Does it deserve its name? When we compare the birthrates of the elite and the disadvantaged, reiterating the stale contrast between Harvard and Harlem, it is natural to wonder how to foster reproductive responsibility. There are two obvious lines of potential action: Abandon the principle of reproductive freedom and starve—or sterilize—poor women into reproductive conformity, or give them the opportunity to make meaningful lives in society, give them a place in the community, so that they too will have interests in shaping the world their children will inhabit. If the second solution cannot be achieved, if pragmatists are right in thinking that the effects of the social lotteries are too hard to undo, then will affluent societies be forced to embrace the first? Must the principle of reproductive freedom—a major line of division between utopian eugenics and its shadier kin—ultimately be compromised? Is it thus too dangerous to introduce new technologies applicable to reproductive choices in situations where citizens may easily come to favor measures to curb what they regard as the irresponsible procreation of the poor?

Utopian eugenics is indeed a fragile enterprise. But critics of the promise of molecular biology are wrong to fear that we shall *inevitably* be swept into repeating past eugenic excesses. As I have tried to show, there are theoretical possibilities, exemplified by utopian eugenics, for a far more benign program. It would be small consolation if that were simply a theoretical point, if the possibilities could not be actualized. Ironically, we can now see that *both* idealist critiques of applications of molecular medicine and pragmatist defenses mix optimism and pessimism. Idealists believe that we can take steps to eradicate the inequalities in affluent democracies, but they are skeptical about the possibility of preventing new injustices spawned by applications of molecular knowledge if the broader social issues are not addressed. Pragmatists maintain that piecemeal efforts (perhaps of the kinds I have sketched in some early chapters) are sufficient to ward off the danger, and that we can sustain a benign form of eugenics, but they doubt our ability to remedy more general social ills. What if both of the pessimistic assessments are correct?

When we broaden our horizon, the worry is reinforced. Throughout this book, I have been focusing on the impact that advances in molecular biology will have on the affluent democracies. Although applications of new technologies have begun in those nations, their impact will not be confined to the first world. The products of biological research have the potential to affect the well-being of people all over the globe.

When resources are limited, we face difficult choices in setting priorities. Restricted appropriations for public health, for human services, and for biomedical research confront us with dilemmas. Should we concentrate on diseases that are traceable to single genes—typically more tractable, often devastating, but affecting a relatively small number of people—or on conditions with complex etiologies that affect millions? Should we attend to genetic causes or try to bring about environmental alterations—large-scale improvements in diet, for example—that could greatly boost the average health of citizens? Should we attack the biological causes of misery or try to remedy the social determinants that depress the quality of lives? Even if we believe, with idealists, that affluent societies have sufficient resources to avoid these dilemmas by redistributing assets, optimism survives only so long as we look within. On a global scale, commitment to the equal well-being of all would confront us with unavoidable hard choices. When millions of children die each year from malaria, how can we justify our expenditures on research into Lesch-Nyhan syndrome, cystic fibrosis, even the common forms of cancer?

Earlier discussions have shown that scale matters. Within a first-world society, problems of genetic disease prove tractable because the number of those afflicted is not too great. But our planet probably already contains too many people. Given the resources available to us, we do not know how to assure everybody—or even a majority—a life of minimally acceptable quality. With advances in medicine, the population problem may only become more severe. Developing and sharing medical technologies, fighting the hostile forces that now check the pace of population growth, seems morally required of us, even though it may add to the numbers of an overcrowded planet. Are we morally committed to courses of action that will utterly debase the lives of our descendants?

Eugenics is inescapable, but the eugenic issues that have occupied us center on the *kinds* of lives we should bring into being. Contemplating the impact of molecular genetics on affluent societies, that focus is entirely appropriate. As we adopt a more global perspective, however, considerations of number can no longer be neglected. We are forced to confront the eugenic question: How many people should there be? But even if a philosopher or political theorist of sublime insight were to deliver a convincing answer to questions about population size, vast practical problems would remain. Would it be possible to keep the world's population beneath the manageable limits prescribed without opposing and overriding religious doctrines and cultural traditions which are fervently upheld? Would it be necessary, in the end, to retract the principle of reproductive freedom that is at the core of utopian eugenics, requiring, as some countries already have, that a woman may not bear more than one or two children? Will our remote progeny look back wistfully on our good fortune in being able to indulge ourselves with respect to family size, or will they see us as irresponsible, sowing the seeds of problems that dominate their world?

The problems of overpopulation and of the global distribution of resources will occupy us and our descendants for any number of pressing reasons. With serious, sustained effort—and with luck—we shall make headway with those problems. Although the challenges are enormous, we may yet be able to foster the ideal that the results of the genetic revolution will become available to all members of our species, that they will transform the quality of human lives across the globe.

• • •

At the end of his life, Sir Isaac Newton, underrating his great accomplishments, characterized himself as "a boy playing on the seashore and diverting myself in now and then finding a smoother pebble or a prettier shell than ordinary, whilst the great ocean of truth lay all undiscovered before me." Forty years on from the double helix, there is a sense today that molecular biology is on the point of launching fully into that great ocean of truth, that coming decades will provide unprecedented—perhaps even now unimaginable—knowledge of how animals work. Among those animals will be our own species, harder though it is to fathom many of our own traits and incomplete though our self-understanding is likely to be. We shall know enough to address many ills that now assail us, to exercise reproductive choices, to foresee and possibly to change our own evolution.

Yet knowledge is dangerous, even when it comes in large draughts. Fear of anticipated advances in molecular genetics stems from awareness that we may expose genetic differences without being able to resolve debilitating genetic conditions, that we shall lengthen our list of prejudices. The deepest motivations for applying the new biology are quite at odds with any such consequences, for the most exhilarating prospect is that we shall be able to repair misfortunes that ground tragic inequalities in the quality of lives. Yet, though genetic diseases may first arouse our compassion, we should reflect further, asking about the scope of the demands upon us. The questions of this chapter are not rhetorical; they are hard problems, demanding sustained thought, research, and public discussion. Just as in common humanity we cannot stand by without concern for the children of Children's, so too we should be moved by the many ways in which mischance is dealt at birth and in the early years. So too we should commit ourselves to ameliorating—insofar as we can—the hideous unfairness of an unequal inheritance.

Postscript (March 1997)

"Researchers Astounded" is not the typical phraseology of a head-line on the front page of the *New York Times* (February 23, 1997). Lamb number 6LL3, better known as Dolly, took the world by sur-prise, sparking debate about the proper uses of biotechnology and inspiring predictable public fantasies (and predictable jokes). Rec-ognizing that what is possible today with sheep will probably be fea-sible with human beings tomorrow, commentators speculated about the legitimacy of cloning Pavarotti or Einstein, about the chances that a demented dictator might produce an army of supersoldiers, and about the future of basketball in a world where the Boston Larry Birds play against the Chicago Michael Jordans. Polls showed that Mother Teresa was the most popular choice for person-to-be-cloned, although a film star (Michelle Pfeiffer) was not far behind, and Bill Clinton and Hillary Clinton obtained some support.

Mary Shelley may have a lot to answer for. The Frankenstein story, typically in one of its film versions, colors popular reception of news about cloning, fomenting a potent brew of associations—we assume that human lives can be created to order, that it can be done instantly, that we can achieve exact replicas, and, of course, that it is all going to turn out disastrously. Reality is much more sobering, and it is a good idea to preface debates about the morality of human cloning with a clear understanding of the scientific facts.

Over two decades ago, a developmental biologist, John Gurdon, reported the possibility of cloning amphibians. Gurdon and his coworkers were able to remove the nucleus from a frog egg and replace it with the nucleus from an embryonic tadpole. The animals survived and developed to adulthood, becoming frogs that had the same complement of genes within the nucleus as the tadpole embryo. Further efforts to use nuclei from adult donors were unsuccessful. When the nucleus from a frog egg was replaced with that from a cell taken from an adult frog, the embryo died at a relatively early stage of development. Moreover, nobody was able to perform Gurdon's original trick on mammals. Would-be cloners who tried to insert nuclei from mouse embryos into mouse eggs consistently ended up with dead fetal mice. So, despite initial hopes and fears, it appeared that the route to cloning adult human beings was doubly blocked: only transfer of embryonic DNA seemed to work, and even that failed in mammals.

Biologists had an explanation for the failure to produce normal development after inserting nuclei from adult cells. Although adult cells contain all the genes, they are also *differentiated,* set to perform particular functions, and this comes about because some genes are expressed in them, while others are "turned off." Regulation of genes is a matter of the attachment of proteins to the DNA so that some regions are accessible for transcription and others are not. So it was assumed that chromosomes in adult cells would have a complex coating of proteins on the DNA and this would prevent the transcription of genes that need to be activated in early development. In consequence, transferring nuclei from adult cells always produced inviable embryos. The "high-tech" solution to the problem—the "obvious" solution from the viewpoint of molecular genetics—is to use the arsenal of molecular techniques to strip away the protein coating, restore the DNA to its (presumed) original condition, and only then transfer the nucleus to the recipient egg. To date, nobody has managed to make this approach work.

The breakthrough came not from one of the major centers in which the genetic revolution is whirling on, but from the far less glamorous world of animal husbandry and agricultural research. In 1996, a team of workers at the Roslin Institute near Edinburgh, led by Dr. Ian Wilmut, announced that they had succeeded in producing two live sheep, Megan and Morag, by transplanting nuclei from embryonic sheep cells. One barrier had been breached: Wilmut and his colleagues had shown that just what Gurdon had done in frogs could be achieved in sheep. Yet it seemed that the major problem, that of tricking an egg into normal development when equipped with an *adult* cell nucleus, still remained. In retrospect, we can recognize that not quite enough attention was given to Wilmut's first announcement, for Megan and Morag testified to a new technique of nuclear transference.

Wilmut conjectured that the failures of normal development resulted from the fact that the cell that supplied the nucleus and the egg that received it were at different stages of the cell cycle. Using well-known techniques from cell biology, he "starved" both cells, so that both were in an inactive phase at the time of transfer. In a series of experiments, he discovered that inserting nuclei from adult cells (from the udder of a pregnant ewe) under this regime gave rise to a number of embryos, which could be implanted in ewes. Although there was a high rate of miscarriage, one of the pregnancies went to term. So, after beginning with 277 successfully transferred nuclei, Wilmut obtained one healthy lamb—the celebrated Dolly.

Wilmut's achievement raises three important questions: Will it be possible to perform the same operations on human cells? Will cloners be able to reduce the high rate of failure? What exactly is the relationship between a clone obtained in this way and previously existing animals? Answers to the first two of these are necessarily tentative, since predicting even the immediate trajectory of biological research is always vulnerable to unforeseen contingencies. (In the weeks after Gurdon's success, it seemed that cloning all kinds of animals was just around the corner; from the middle 1980s to 1996, it appeared that cloning adult mammals was a science-fiction fantasy.) However, unless there is some quite unanticipated snag, we can expect that Wilmut's technique will *eventually* work just as well on human cells as it does in sheep, and that failure rates in sheep (or in other mammals) will quickly be reduced.

Assuming that Wilmut's diagnosis of the problems of mammalian cloning is roughly correct, then the crucial step involves preparing the cells for nuclear transfer by making them quiescent. Of course, learning how to "starve" human cells so that they are ready may involve some experimental tinkering. Probably there would be a fair bit of trial-and-error work before the techniques became sufficiently precise to allow for embryos to develop to the stage at which they can be implanted with a very high rate of success, and to overcome any potential difficulties with implantation or with the resultant pregnancy. Many of the problems that prospective human cloners would face are likely to be analogues of obstacles to the various forms of assisted reproduction—and it is perfectly possible that the successes of past human reproductive technology would smooth the way for cloning.

On the third question we can be more confident. Dolly has the same nuclear genetic material as the adult pregnant ewe, from whose udder cell the inserted nucleus originally came. A different female supplied the egg into which the nucleus was inserted, and Dolly thus has the same mitochondrial DNA as this ewe; indeed, her early development was shaped by the interaction between the DNA in the nucleus and the contents of the cytoplasm, the contributions of different adult females. Yet a third sheep, the ewe into which the embryonic Dolly was implanted, played a role in Dolly's nascent life, providing her with a uterine environment. In an obvious sense, Dolly has three mothers—nucleus mother, egg mother, and womb mother—and no father (unless, of course, we give Dr. Wilmut that honor for his guiding role).

Now imagine Polly, a human counterpart of Dolly. Will Polly be a replica of any existing human being? Certainly she will not be the same person as any of the mothers—even the nuclear mother. Personal identity, as philosophers since John Locke have recognized, depends on continuity of memory and other psychological attitudes. There is no hope of ensuring personal survival by arranging for cloning through supplying a cell nucleus. Megalomaniacs with intimations of immortality need not apply.

Yet you might think that Polly might be very similar to her nuclear mother, perhaps extremely similar if we arranged for the nuclear mother to be the same person as the egg mother, and for that person's mother to be the womb mother. That combination of "parents" would seem to turn Polly into a close approximation of her nuclear mother's identical twin. An approximation, perhaps, but nobody knows how close. Polly and her nuclear mother differ in three ways in which identical twins are typically the same. They develop from eggs with different cytoplasmic constitutions, they are not carried to term in a common uterine environment, and their environments after birth are likely to be quite different.

Interestingly, during the next few years, Wilmut's technique will allow us to remedy our ignorance about the relative importance of various causes of phenotypic traits by performing experiments on nonhuman mammals. It will be possible to develop organisms with the same nuclear genes within recipient eggs with varied cytoplasms. By exploring the results, biologists will be able to discover the extent to which constituents of the egg outside the nucleus play a role in shaping the phenotype. They will also be able to explore the ways in which the uterine environment makes a difference. Perhaps they will find that variation in cytoplasm and difference in womb have little effect, in which case Polly will be a better approximation to an identical twin of her nuclear mother. More probably, I believe, they will expose some aspects of the phenotype that are influenced by the character of the cytoplasm or by the state of the womb, thus identifying ways in which Polly would fall short of perfect twinhood.

Even before these experiments are done, we know of some important differences between Polly and her nuclear mother. Unlike most identical twins, they will grow up in environments that are quite dissimilar, if only because the gap in their ages will correspond to changes in dietary fashions, educational trends, adolescent culture, and so forth. When these sources of variation are combined with the more uncertain judgments about effects of cytoplasmic factors and the prenatal environment, we can conclude that human clones will be less alike than identical twins, and quite possibly very much less alike.

Those beguiled by genetalk move quickly from the idea that clones are genetically identical (which is, to a first approximation, correct) to the view that clones will be replicas of one another. Identical twins reared together are obviously similar in many respects, but they are by no means interchangeable people. It is pertinent to recall the statistics about sexual orientation: 50 percent of male (identical) twins who are gay have a co-twin who is not. Minute differences in shared environments can obviously play a large role. How much more dissimilarity can we anticipate given the much more dramatic variations that I have indicated?

There will never be another you. If you hoped to fashion a son or daughter exactly in your own image, you would be doomed to disappointment. Nonetheless, you might hope to use cloning technology to have a child of a particular kind—just as the obvious agricultural applications focus on single features of domestic animals, like their capacity for producing milk. Some human characteristics are under tight genetic control, and if we wanted to ensure that our children carried genetic diseases like Huntington's and Tay-Sachs, then, of course, we could do so, although the idea is so monstrous that it only surfaces in order to be dismissed. Perhaps there are other features that are relatively insusceptible to niceties of the environment, aspects of body morphology, for example. An obvious example is eye color. Imagine a couple determined to do what they can to produce a

Hollywood star. Fascinated by the color of Elizabeth Taylor's eyes, they obtain a sample of tissue from the actress and clone a young Liz. For reasons already discussed, it is probable that Elizabeth II would be different from Elizabeth I, but we might think that she would have that distinctive eye color. Supposing that to be so, should we conclude that the couple will realize their dream? Probably not. Waiving issues about intelligence, poise, and acting ability and supposing that the movie moguls of the future respond only to physical attractiveness, the eyes may not have it. Apparently tiny incidents in early development may modify the shape of the orbits, producing a combination of features in which the eye color no longer has its bewitching effect. At best, the confused couple can only hope to raise the probability that their daughter will capture the hearts of millions.

Physical attractiveness, the real target of the couple's plan, turns on more than eye color, and that is the general way of things. The traits we value most are produced by a complex interaction between genotypes and environments, and by fixing the genotype, we can only increase our chances of achieving the results we want. Demented dictators bent on invading their neighbors can do no more than add to the likelihood of generating the "master race." Before we startle ourselves with the imagined sound of jackboots marching across the frontier, we should remember that there is no shortcut to the process of rearing children and training them in whatever ways are appropriate to our ends. Indeed, when we appreciate that point, we can see that if the dictators are slightly less demented, they will do what military recruiters have always done, namely select on grounds of physical fitness, ease of indoctrination, courage, and such traits and then invest extensively in military academies. Cloning adds very little to the chances of success.

Similar points apply to the fantasies of cloning Einstein, Mother Teresa, or Yo-Yo Ma. The chances of generating true distinction in any area of complex human activity, whether it be scientific accomplishment, dedication to the well-being of others, or artistic expression, are infinitesimal. *Perhaps* cloning would allow us to raise the probability from infinitesimal to very, very tiny. A program designed to use cloning to transform human life by having a higher number of outstanding individuals would, at most, give a minute number of "successes" at the cost of vastly more "failures." Those who worry that Dolly is one survivor among 277 attempts should find this scenario far more disturbing.

Garish popular fantasies dissolve when confronted with the facts about the biotechnology of cloning, suggesting that only rich recluses, hermetically sealed in ignorance, should be tempted by the projects that fascinate and horrify us most. Yet there are other more mundane ventures that have a closer connection with reality. Parents who demand less than truly outstanding performance, but still have a preferred dimension on which they want their children to excel, might turn to cloning in hopes of raising their chances. Had my wife and I been seriously concerned to bring into the world sons who would have dominated the basketball court or been mainstays of the defensive line, then we would have been ill-advised to proceed in the old-fashioned method of reproduction. At a combined weight of less than 275 pounds and a combined height of just over eleven feet, we would have done far better to transfer a nucleus from some strapping star of the NBA or the NFL. Perhaps, by doing so, we would significantly have raised the chances of having a son on the high-school basketball or football team. Success, even at that rather modest goal, would not have been guaranteed, since there are all kinds of ways in which the boy's development might have gone differently (think of accidents, competing interests, dislike of competitive sports, and so forth). Nor would cloning necessarily have been the best way for us to proceed: Maybe we could have employed the results of genetic testing to produce, by *in vitro* fertilization, a fertilized egg having alleles at crucial loci that predispose to a large, muscular body; maybe we could have used artificial insemination, or have adopted a son. Nevertheless, cloning would surely have raised the probabilities of our obtaining the child we wanted.

Just that final phrase indicates the moral squalor of the story. As I have imagined it, we have a plan for the life to come laid down in advance; we are determined to do what we can to make it come out a certain way—and, presumably, if it does not come out that way, it will count as a failure. Throughout the discussion of utopian eugenics, I insisted that prenatal decisions should be guided by reflection on the quality of the nascent life, and I understood that in terms of creating the conditions under which a child could form a central set of desires, a conception of what his or her life means that had a decent chance of being satisfied. In the present scenario, there is a crass failure to recognize the child as an independent being, one who should form his own sense of who he is and what his life means. The contours of the life are imposed from without.

Parents have been tempted to do similar things before. James Mill had a plan for his son's life, leading him to begin young John Stuart's instruction in Greek at age three and his Latin at age eight. John Stuart Mill's *Autobiography* is a quietly moving testament to the cramping effect of his felt need to live out the life his eminent father had designed for him. In early adulthood, Mill *fils* suffered a nervous breakdown, from which he recovered, going on to a career of great intellectual distinction. Although John Stuart partially fulfilled his father's aspirations for him, one of the most striking features of his philosophical work is his passionate defense of human freedom. The central point about what was wrong with this father's attitude toward a son has never been better expressed than in the splendid prose of *On Liberty:* "Mankind are greater gainers by suffering each other to live as seems good to themselves than by compelling each to live as seems good to the rest."

If cloning human beings is undertaken in the hope of generating a particular kind of person, a person whose standards of what matters in life are imposed from without, then it is morally repugnant, not because it involves biological tinkering but because it is continuous with other ways of interfering with human autonomy that we ought to resist. Human cloning would provide new ways of committing old moral errors. To discover whether or not there are morally permissible cases of cloning, we need to see if this objectionable feature can be removed, if there are situations in which the intention of the prospective parents is properly focused on the quality of human lives but in which cloning represents the only option for them. Three scenarios come immediately to mind.

The case of the dying child. Imagine a couple whose only son is slowly dying. If the child were provided with a kidney transplant within the next ten years, he would recover and be able to lead a normal life. Unfortunately, neither parent is able to supply a compatible organ, and it is known that individuals with kidneys that could be successfully transplanted are extremely rare. However, if a brother were produced by cloning, then it would be possible to use one of his kidneys to save the life of the elder son. Supposing that the technology of cloning human beings has become sufficiently reliable to give the couple a very high probability of successfully producing a son with the same complement of nuclear genes, is it permissible for them to do so?

The case of the grieving widow. A woman's much-loved husband has been killed in a car crash. As the result of the same crash, the couple's only daughter lies in a coma, with irreversible brain damage, and she will surely die in a matter of months. The widow is no longer able to bear children. Should she be allowed to have the nuclear DNA from one of her daughter's cells inserted in an egg supplied by another woman, and to have a clone of her child produced through surrogate motherhood?

The case of the loving lesbians. A lesbian couple, devoted to one another for many years, wish to produce a child. Because they would like the child to be biologically connected to each of them, they request that a cell nucleus from one of them be inserted in an egg from the other, and that the embryo be implanted in the woman who donated the egg. (Here, one of the women would be nuclear mother and the other would be both egg mother and womb mother.) Should their request be accepted?

In all of these instances, unlike the ones considered earlier, there is no blatant attempt to impose the plan of a new life, to interfere with a child's own conception of what is valuable. Yet there are lingering concerns that need to be addressed. The first scenario, and to a lesser extent the second, arouses suspicion that children are being subordinated to special adult purposes and projects. Turning from John Stuart Mill to one of the other great influences on contemporary moral theory, Immanuel Kant, we can formulate the worry as a different question about respecting the autonomy of the child: Can these cases be reconciled with the injunction "to treat humanity whether in your own person or in that of another, always as an end and never as a means only"?

It is quite possible that the parents in the case of the dying child would have intentions that flout that principle. They have no desire for another child. They are desperate to save the son they have, and if they could only find an appropriate organ to transplant, they would be delighted to do that—for them the younger brother would simply be a cache of resources, something to be used in saving the really important life. Presumably, if the brother were born and the transplant did not succeed, they would regard that as a failure. Yet the parental attitudes do not have to be so stark and callous (and, in the instances in which parents have actually contemplated bearing a child to save an older sibling, it is quite clear that they are much more complex). Suppose we imagine that the parents plan to have another child in any case, that they are committed to loving and cherishing the child for his or her own sake. What can be the harm in planning that child's birth so as to allow their firstborn to live?

The moral quality of what is done plainly depends on the parental attitudes, specifically on whether or not they have the proper concern for the younger boy's well-being, independently of his being able to save his elder brother. Ironically, their love for him may be manifested most clearly if the project goes awry and the first child dies. Although that love might equally be present in cases where the elder son survives, reflective parents will probably always wonder whether it is untinged by the desire to find some means of saving the first-born—and, of course, the younger boy is likely to entertain worries of a similar nature. He would by no means be the first child to feel himself a second-class substitute, in this case either a helpmeet or a possible replacement for someone loved in his own right.

Similarly, the grieving widow might be motivated solely by desire to forge some link with the happy past, so that the child produced by cloning would be valuable because she was genetically close to the dead (having the same nuclear DNA as her sister, DNA that derives from the widow and her dead husband). If so, another person is being treated as a means to understandable, but morbid, ends. On the other hand, perhaps the widow is primarily moved by the desire for another child, and the prospect of cloning simply reflects the common attitude of many (though not all) parents who prize biological connection to their offspring. However, as in the case of the dying child, the participants, if they are at all reflective, are bound to wonder about the mixture of attitudes surrounding the production of a life so intimately connected to the past.

The case of the loving lesbians is the purest of the three, for here we seem to have a precise analogue of the situation in which heterosexual couples find themselves. Cloning would enable the devoted pair to have a child biologically related to both of them. There is no question of imposing some particular plan on the nascent life, even the minimal one of hoping to save another child or to serve as a reminder of the dead, but simply the wish to have a child who is their own, the expression of their mutual love. If human cloning is ever defensible, it will be in contexts like this.

During past decades, medicine has allowed many couples to overcome reproductive problems and to have biological children. The development of techniques of assisted reproduction responds to the sense that couples who have problems with infertility have been deprived of something that it is quite reasonable for people to value, and that various kinds of manipulations with human cells are legitimate responses to their frustrations. Yet serious issues remain. How close an approximation to the normal circumstances of reproduction and the normal genetic connections should we strive to achieve? How should the benefits of restoring reproduction be weighed against possible risks of the techniques? Both kinds of questions arise with respect to our scenarios.

Lesbian couples already have an option to produce a child who will be biologically related to both. If an egg from one of them is fertilized with sperm (supplied, say, by a male relative of the other) and the resultant embryo is implanted in the womb of the woman who did not give the egg, then both have a biological connection to the child (one is egg mother, the other womb mother). That method of reproduction might even seem preferable, diminishing any sense of burden that the child might feel because of special biological closeness to one of the mothers and allowing for the possibility of having children of either sex. The grieving widow might turn to existing techniques of assisted reproduction and rear a child conceived from artificial insemination of one of her daughter's eggs. In either case, cloning would create a closer biological connection—but should that extra degree of relationship be assigned particularly high value?

My discussion of all three scenarios also depends on assuming that human cloning works smoothly, that there are no worrisome risks that the pregnancy will go awry, producing a child whose development is seriously disrupted. Dolly, remember, was one success out of 277 tries, and we can suppose that early ventures in human cloning would have an appreciable rate of failure. We cannot know yet whether the development of technology for cloning human beings would simply involve the death of early embryos, or whether, along the way, researchers would generate malformed fetuses and, from time to time, children with problems undetectable before birth. During the next few years, we shall certainly come to know much more about the biological processes involved in cloning mammals, and the information we acquire may make it possible to undertake human cloning with confidence that any breakdowns will occur early in development (before there is a person with rights). Meanwhile, we can hope that the continuing transformation of our genetic knowledge will provide improved methods of transplantation (see page 112), and thus bring relief to parents whose children die for lack of compatible organs.

Should human cloning be banned? Until we have much more extensive and detailed knowledge of how cloning can be achieved (and what the potential problems are) in a variety of mammalian species, there is no warrant for trying to perform Wilmut's clever trick on ourselves. I have suggested that there are some few circumstances in which human cloning might be morally permissible, but, in at least two of these, there are genuine concerns about attitudes to the nascent life, while, in the third, alternative techniques, already available, offer almost as good a response to the underlying predicament. Perhaps, when cloning techniques have become routine in nonhuman mammalian biology, we may acknowledge human cloning as appropriate relief for the parents of dying children, for grieving widows, and for loving lesbians. For now, however, we do best to try to help them in other ways.

Dolly, we are told, like the scientist who helped her into existence, is learning to live with the television cameras. Media fascination with cloning plainly reached the White House, provoking President Clinton first to refer the issue to his newly formed Bioethics Advisory Committee, later to ban federal funding of applications of cloning technology to human beings. The February 27, 1997, issue of *Nature* featuring Dolly offered a less-than-positive assessment of the presidential reaction: "At a time when the science policy world is replete with technology foresight exercises, for a U.S. president and other politicians only now to be requesting guidance about what appears in today's *Nature* is shaming."

At the first stages of the Human Genome Project, James Watson argued for the assignment of funds to study the "ethical, legal, and social implications" of the purely scientific research. Watson explicitly drew the analogy with the original development of nuclear technology, recommending that, this time, scientific and social change might go hand in hand. Almost a decade later, the mapping and sequencing are advancing faster than most people had anticipated—and the affluent nations remain almost where they were in terms of supplying the social backdrop that will put the genetic knowledge to proper use. That is not for lack of numerous expert studies that outline the potential problems and that propose ways of overcoming them. Much has been written. Little has been done. In the United States we still lack the most basic means of averting genetic discrimination, to wit universal health coverage (see page 136), but Britain and even the continental European nations are little better placed to cope with what is coming.

The moral problems raised by the possibility of human cloning should be addressed by drawing on general moral principles, articulated in many contexts and in many idioms in the history of thought. (Throughout this book, I have tried to formulate them in the context of our rapidly advancing biological knowledge.) Those principles advance a conception of what matters in human life, of what proper attitudes toward others should be, and, more specifically, of how we should treat nascent lives. Once the factual confusions about cloning are cleared away and once we have appealed to broader moral theory, we can see how to navigate our way through the territory and which possibilities are especially difficult. However, the moral principles cannot be applied selectively, nor can we dodge their implications either within the domain of uses of biotechnology or in thinking about our duties to children who are currently deprived of genuine opportunities for health and happiness. Only moral chameleons call for a ban on human cloning because of remote potential harms while instituting or supporting policies that permit children to live without proper health care and even endanger their prospects of food and shelter.

The belated response to cloning is of a piece with a general failure to translate clear moral directives into regulations and policies.

Dolly is a highly visible symbol, but behind her is a broad array of moral issues that citizens of affluent societies seem to prefer to leave in the shadows. However strongly we feel about the plight of loving lesbians, grieving widows, or even couples whose children are dying, deciding the legitimate employment of human cloning in dealing with their troubles is not our most urgent problem. Those who think that working out the proper limits of human cloning is the big issue are suffering from moral myopia.

General moral principles provide us with an obligation to improve the quality of human lives, where we have the opportunity to do so, and developments in biotechnology provide opportunities and challenges. If we took the principles seriously, we would be led to demand serious investment in programs to improve the lives of the young, the disabled, and the socially disadvantaged. That is not quite what is going on in the "civilized" world. Making demands for social investment seems quixotic, especially at a time when, in America, funds for poor children and disabled people who are out of work are being slashed, and when, in other affluent countries, there is serious questioning of the responsibilities of societies to their citizens. Yet the application of patronizing adjectives does nothing to undermine the legitimacy of the demands. What is truly shameful is not that the response to possibilities of cloning came so late, nor that it has been confused, but the common reluctance of all the affluent nations to think through the implications of time-honored moral principles and to design a coherent use of the new genetic information and technology for human well-being.

—Philip Kitcher

Notes

CHAPTER 1   *The Shapes of Suffering*

15   *Others think differently:* A sustained presentation of worries about genetic testing is offered in Ruth Hubbard and Elijah Wald, *Exploding the Gene Myth* (Boston: Beacon, 1993). In *Dangerous Diagnostics,* 2d ed. (Chicago: University of Chicago Press, 1994) Dorothy Nelkin and Laurence Tancredi make similar points in a more muted and nuanced way.

CHAPTER 2   *Our Mortal Coils*

23   *the results of the analysis, the "genetic report card":* Many champions of molecular biology and its impact on the medicine of the next century have suggested this kind of development in pediatric practice. Francis Collins, current director of Human Genome Research at the (American) National Institute of Health, has frequently talked about the "genetic report card" in describing the goals of the Human Genome Project.

25   *(such as Tay-Sachs disease among Jews of Ashkenazi descent and thalassemia among Cypriots and other Mediterranean people):* For the story of Tay-Sachs testing, see M. Kaback et al., "Tay-Sachs Disease: A Model for the Control of Recessive Genetic Disorders," in *Birth Defects,* eds. A. Motulsky and W. Lenz (Amsterdam: Excerpta Medica, 1986), pp. 248–262. An illuminating discussion of thalassemia testing in Cyprus is M. Angastiniotis et al., "How Thalassemia was Controlled in Cyprus," *World Health Forum* 7 (1986): 291–297.

26   *to bring into clearer focus the great achievements of late-twentieth-century molecular biology:* There are now many excellent books that will supplement the whirlwind tour of the present chapter and give lucid expositions of the scientific

details. Christopher Wills's *Exons, Introns, and Talking Genes* (New York: Basic Books, 1991) is extremely accessible, leisurely, and avuncular. Paul Berg and Maxine Singer, *Dealing with Genes* (Mill Valley, Calif.: University Science Books, 1992), demands more of the reader but offers a superb account of recent molecular biology, with beautiful illustrations. Finally, for those who want serious detail, there is no better source than James D. Watson et al., *Recombinant DNA,* 2d ed. (New York: W. H. Freeman, 1992), which is a pedagogical masterpiece.

28   *The course of this disease is invariable:* In presenting examples of genetic diseases that are traceable to mutations at a single locus, I have relied upon Kenneth Jones, *Smith's Recognizable Patterns of Human Malformation,* 4th ed. (Philadelphia: W. B. Saunders, 1988), and J. M. Connor and M. A. Ferguson-Smith, *Essential Medical Genetics,* 4th ed. (Oxford: Blackwell, 1993).

29   *the two young authors, James Watson and Francis Crick:* Watson's own account of the discovery of the structure of DNA is given in *The Double Helix* (New York: Norton, 1967). For a magisterial study of the discovery and its impact during the next two decades, the reader should consult Horace Freeland Judson, *The Eighth Day of Creation* (New York: Simon & Schuster, 1979). Although Robert Cook-Deegan (*The Gene Wars* [New York: Norton, 1994]) provides a comprehensive and informative account of the early career of the Human Genome Project, the history of molecular biology from 1970 to the present has still to be written.

31   *"It has not escaped our notice":* J. D. Watson and F. Crick, "A Structure for Deoxyribose Nucleic Acid," *Nature* (April 25, 1953): 737. A facsimile of the paper is reproduced in the Norton Critical edition of *The Double Helix,* p. 238 (see also p. 240 for the printed text).

36   *Doctors have discovered:* The discovery of the mutation responsible for defective Surfactant B proteins, and consequent respiratory failure, is reported in L. M. Nogee et al., "A Mutation in the Surfactant Protein B Gene Responsible for Fatal Respiratory Disease in Multiple Kindreds," *Journal of Clinical Investigation* 93 (1994): 1860–1863.

39   *the dedication and energy that Nancy Wexler:* There are many available accounts of the identification of the gene whose mutation results in Huntington's disease. Nancy Wexler's own description of the research is given in "Clairvoyance and Caution: Repercussions from the Human Genome Project" in *The Code of Codes,* eds. D. Kevles and L. Hood (Cambridge, Mass.: Harvard, 1992), pp. 211–243. Wexler uses the story to draw several morals of the kind I emphasize in subsequent chapters.

57   *The most common cause of mental retardation, Fragile X syndrome:* A thorough review of the syndrome and of the underlying genetics is provided in R. Hagerman and A. Silverman, *Fragile X Syndrome: Diagnosis, Treatment, and Research* (Baltimore: Johns Hopkins, 1991).

61   *These complexities are illustrated:* For an excellent discussion of the molecular biology of various kinds of cancer, see J. D. Watson et al., *Recombinant DNA,* chap. 18. R. C. Williams Jr., *Molecular Biology in Clinical Medicine* (New York: Elsevier, 1991) is also a valuable source.

62 *genes for colon cancer and early-onset breast cancer:* The sequencing of the much-sought breast cancer gene, *BRCA1*, is reported in Y. Miki et al., "A Strong Candidate for the Breast and Ovarian Cancer Susceptibility Gene *BRCA1*," *Science* 266 (1994): 66–71.

62 *Only about 10 percent of those affected with hypercholesterolemia:* For a more detailed discussion, see N. A. Holtzman, *Proceed with Caution* (Baltimore: Johns Hopkins, 1989), pp. 22–23, 44.

CHAPTER 3   *To Test or Not to Test?*

66 *Children who carry two copies of a recessive allele:* The molecular basis of PKU is lucidly described by W. L. Nyhan in "Approaches to the Treatment of Inherited Metabolic Disease" (chapter 1 of *Therapy for Genetic Disease,* ed. T. Friedmann [Oxford: Oxford University Press, 1991]).

66 *the historian of biology Diane Paul has shown:* An outline of Paul's research is given in her "Genetic Screening: Lessons from the History of PKU," to appear in M. Forbes et al., *Proceedings of the Philosophy of Science Association 1994* (Chicago: University of Chicago Press, 1995). I am extremely grateful to Paul for sharing her unpublished work with me.

67 *The PKU diet is both unpleasant and expensive:* Sample diets for three patients of different ages are given in E. Nichols, *Human Gene Therapy* (Cambridge, Mass.: Harvard, 1988), p. 71. The basis of the diet is always a mixture of commercially prepared formula with a few additives. (I have been told that the current cost of the formula is approximately four thousand dollars per year for one person). People on the diet are also allowed to eat small quantities of bread, fruit, potatoes, cookies, and syrup. Most of the staples of childhood and teenage eating are taboo.

67 *they feel themselves inferior to their contemporaries who can eat normally:* For some insights into the attitudes of children and adolescents on the PKU diet, see Joan Moen et al., "PKU as a Factor in the Development of Self-esteem," *Journal of Pediatrics* 90 (1977): 1027–1029.

68 *that is precisely what we want to know:* In an enormous and growing number of cases, discoveries in human molecular genetics advance our power to diagnose diseases. These advances tend to be underrated because of our fascination with prediction. In public presentations, Eric Lander has emphasized the diagnostic power of the new molecular genetics.

69 *Our tendency to confound probabilities:* A penetrating account of the XYY fallacy is given by J. Beckwith and J. King, "The XYY Syndrome: A Dangerous Myth," *New Scientist* 64 (1974): 474–476. See also Beckwith, "A Historical View of Social Responsibility in Genetics," *BioScience* 43 (1993): 327–333.

69 *It is possible, although researchers think it highly unlikely:* Most of those working on Fragile X syndrome believe that the indirect estimates obtained from genotype-phenotype comparisons within families in which someone has ante-

cedently been diagnosed with the syndrome show that males with the "full mutation" (more than three hundred copies of the repeat) will always suffer retardation and the behavioral effects. They can point to a study of 318 Fragile X families which shows a 100 percent rate of mental retardation for men with the full mutation (F. Rousseau et al., "A Multi-Center Study on Genotype-Phenotype Correlations in the Fragile X Syndrome, Using Direct Diagnosis with Probe StB12.3: The First 2,253 Cases," *American Journal of Human Genetics* 55 [1994]: 225–237). This study, and the other investigations that confine themselves to families in which patients have already been diagnosed, do not eliminate the logical possibility that the groups of people studied have some further common factor, genetic or environmental, that interacts with the long repeat to produce the phenotype. (I am grateful to Dr. G. Sutherland and Dr. J. Witkowski for information on the investigation of genotype-phenotype correlations in Fragile X; they believe that the existing studies are conclusive; I suggest that the logical possibility should not be discounted, especially when it is so easy to investigate the asymptomatic population in the way described in the text.)

71 *The parents would be quite misled:* Various authors have warned against leaping too quickly to conclusions about phenotypes. See, for example, Dorothy Nelkin and Laurence Tancredi, *Dangerous Diagnostics,* 2d ed. (Chicago: University of Chicago Press, 1994) and N. A. Holtzman, *Proceed with Caution* (Baltimore: Johns Hopkins, 1989).

73 *Even when advice about possible changes in habits:* This point is made forcefully in Ruth Hubbard and Elijah Wald, *Exploding the Gene Myth* (Boston: Beacon, 1993).

74 *possible side effects of mammograms:* Worries about the safety of mammograms are based upon the idea that even small doses of radiation may trigger molecular changes that begin tumor formation. As this book was in press, scientists identified an allele implicated in a neurological disorder (when the allele is present in double dose) that seems to confer upon women who carry one copy a predisposition to breast cancer. It has been suggested that cells carrying one such allele might be less efficient at DNA repair, so that women with one copy might actually be exposed to greater risks through mammograms. Until the details have been worked out much more carefully, this is highly speculative, but it is theoretically possible that some women who are at greater risk for breast cancer might best be advised *not* to have mammograms—and that future genetic testing may be able to tell them that.

74 *Within those families with a history of breast and ovarian cancer:* For an extremely sensitive discussion of the role of genetic testing in such families, see M-C. King et al., "Inherited Breast and Ovarian Cancer: What Are the Risks? What Are the Choices?," *Journal of the American Medical Association* 269 (1993) 1975–1980.

75 *Tests may even become institutionalized in schools:* There are extremely serious questions, which I do not have the space to pursue in the detail they deserve, about who makes decisions about testing children and adolescents, when young

subjects should be tested, and what exactly they should be told. For preliminary forays into these topics, see Dorothy Wertz et al., "Genetic Testing for Children and Adolescents: Who Decides?," *Journal of the American Medical Association* 272 (1994) 875–881.

76 *Since the arrival of tests for Huntington's disease:* The disparity between the rates of those who predicted they would want to know and those who actually decide to be tested is described in Nancy Wexler, "Clairvoyance and Caution: Repercussions from the Human Genome Project" (in *The Code of Codes,* eds. D. Kevles and L. Hood [Cambridge, Mass.: Harvard, 1992], pp. 211–243), and in L. B. Andrews et al., *Assessing Genetic Risk* (Washington, D.C.: National Academy Press, 1994), pp. 88–89.

76 *Some people know that they would be crushed by bad news:* In a study reported by Neil A. Holtzman (*Proceed with Caution,* 180), 13 of 87 people willing to be tested for Huntington's disease declared that they would commit suicide if the results showed they had the disease.

77 *If only because all citizens pay for the research:* The need to ensure that members of all groups and strata of society benefit from genetic testing has been eloquently expressed by Troy Duster in *Backdoor to Eugenics* (New York: Routledge, 1990), pp. 64–66.

77 *Members of ethnic groups most at risk for stigmatization:* There is a growing literature concerned with the transmission of biomedical information to members of different ethnic groups. For sensitive portraits of serious difficulties and miscommunications, see Barbara Katz Rothman, *The Tentative Pregnancy* (New York: Viking, 1986); Rayna Rapp, "Amniocentesis in Socio-cultural Perspective" (*Journal of Genetic Counselling* 2 [1994]: 183–196); and Patricia King, "The Past as Prologue: Race, Class, and Gene Discrimination,*"* in *Gene Mapping: Using Law and Ethics as Guides,* eds. G. Annas and S. Elias (New York: Oxford, 1992), pp. 94–111.

79 *Critics of genetic testing:* See, in particular, Ruth Hubbard and Elijah Wald, *Exploding the Gene Myth.*

83 *The tendency, in many discussions of moral issues surrounding pregnancy:* For a more extensive treatment of some of the points made in this paragraph, see Barbara Katz Rothman, *The Tentative Pregnancy,* and Ruth Schwartz Cowan, "Genetic Technology and Reproductive Choice: An Ethics for Autonomy" (in *The Code of Codes,* eds. D. Kevles and L. Hood [Cambridge, Mass.: Harvard, 1992], pp. 244–263.)

84 *there is an identifiable individual to whom irreversible damage would be done:* Here I foreshadow the treatment of abortion that will appear later in the text. For a similar perspective on these issues, see R. Dworkin, *Life's Dominion* (New York: Knopf, 1993).

85 *Disability activists have been vocal:* See, for example, Marsha Saxton, "Prenatal Screening and Discriminatory Attitudes About Disability" in *Embryos, Ethics, and Women's Rights,* ed. E. Hoffman Baruch et al. (New York: Harrington Park Press, 1988), pp. 217–224.

85   *In Cyprus, the Greek Orthodox Church:* The success of the program is described by M. Angastiniotis et al., "How Thalassemia Was Controlled in Cyprus," *World Health Forum* 7 (1986): 291–297.

CHAPTER 4   *The Road to Health?*

88   *No less a figure than Walter Gilbert:* See W. Gilbert, "A Vision of the Grail," in *The Code of Codes,* eds. D. Kevles and L. Hood (Cambridge, Mass.: Harvard, 1992), pp. 83–97.

88   *Promises, promises:* For a trenchant essay contending that the promise of the Human Genome Project has been overblown, see R. C. Lewontin, "The Dream of the Human Genome," in his *Biology and Ideology* (New York: Harper, 1992), pp. 61–83.

89   *They are confident of knowing:* For the details of the knowledge obtained by sequencing a yeast chromosome and the current state of yeast genome sequencing, see M. Johnston et al., "Complete Nucleotide Sequence of *Saccharomyces cerevisiae* Chromosome VIII," *Science* 265 (1994): 2077–2082.

90   *But we are on the verge:* A major international effort produced the unprecedented 2.2 Mb of nematode sequence, announced in "2.2 Mb of Contiguous Nucleotide Sequence from Chromosome III of *C. elegans,*" *Nature* 368 (1994): 32–38.

90   *In the last decades, Drosophila melanogaster has assumed new importance:* A lucid review of the major breakthroughs in the molecular genetics of fruit flies is Peter A. Lawrence, *The Making of a Fly* (Oxford: Blackwell, 1992).

91   *Enthusiasts sometimes envisage a royal road:* This position is most fully developed by Walter Gilbert in "A Vision of the Grail."

92   *Patience counsels them to sort through the dross:* Some researchers, notably Craig Venter, believe that there is a shortcut. By picking out mRNAs from human cells and using the enzyme reverse transcriptase it is possible to manufacture a cDNA that corresponds to the exons of the pertinent gene. Using a variety of tricks, biologists can then probe the human genome and locate the full gene (including the introns). However, the success of this strategy as a means of finding *all* the genes depends on being able to find *all* the mRNAs. Since we have no way of systematically investigating all human cells at all stages of development, there is good reason to think that the resultant catalog would be incomplete. Nevertheless, Venter's tools may prove extremely valuable if the human genome turns out to be too massive for full sequencing.

99   *but it will not reveal the essence of our humanity:* Gilbert has suggested that the Human Genome Project will answer the question, What makes us human?

100   *Directors of genome centers presently project:* The figures in this paragraph were obtained from interviews with personnel at several genome centers in the United States in 1993–94. I suspect that some of these people might now give more optimistic estimates, because they believe that a breakthrough is in sight.

100   *Of course, molecular biologists anticipate:* As this book was in press, their hopes were given new support by Craig Venter's successful execution of a new strategy

for genomic sequencing about which many molecular geneticists had been skeptical. Venter was able to produce the full sequence of the bacterium *Haemophilus influenzae* Rd ("Whole-Genome Random Sequencing and Assembly of *Haemophilus influenzae* Rd," *Science* 269 [1995]: 496–512) in a far quicker time than had been anticipated, by breaking the bacterial DNA into random fragments, sequencing them, and using a very sophisticated computer program to assemble the pieces. It has been suggested that a similar strategy might make the sequencing of large genomes far more tractable (see *Science* 269 [1995]: 469), and that it might enable us to achieve the full human sequence within five or six years. So *perhaps* the needed breakthrough has already occurred. However, as noted in the text, we should always be aware that scale matters.

101 *Some put their trust in "Star Wars technology":* For an accessible introduction to some methods that used to seem quite promising, see Christopher Wills, *Exons, Introns, and Talking Genes* (New York: Basic Books, 1991), chap. 7.

CHAPTER 5   *A Patchwork of Therapies*

108 *Our bodily machinery:* Excellent overviews of contemporary molecular medicine can be found in R. C. Williams Jr., *Molecular Biology in Clinical Medicine* (New York: Elsevier, 1991); T. Friedmann ed., *Therapy for Genetic Disease* (Oxford: Oxford University Press, 1991); M. Steinberg et al., *Recombinant DNA Technology: Concepts and Biomedical Applications* (Englewood Cliffs: Prentice-Hall, 1993); and E. Nichols, *Human Gene Therapy* (Cambridge, Mass.: Harvard, 1988).

110 *treatments for Gaucher's disease and Wilson's disease:* J. A. Schneider provides a lucid account of treatments for Wilson's disease in chap. 3 of T. Friedmann ed., *Therapy for Genetic Disease.* Similarly helpful is the discussion of therapy for Gaucher's disease in chap. 3 of M. Steinberg et al., *Recombinant DNA Technology: Concepts and Biomedical Applications.* For a new approach inspired by recent advances in recombinant technology, see L. Xu et al., "Correction of the enzyme deficiency in hematopoietic cells of Gaucher patients using a clinically acceptable retroviral supernatant transfection," *Experimental Hematology* 22 (1994): 223–230.

111 *Discovering that the viscosity:* The attempt to treat CF patients by sending an enzyme to clear away unwanted DNA is reported in H. J. Fuchs et al., "Effect of aerosolized recombinant human DNAase on exacerbations of respiratory symptoms and on pulmonary function in patients with cystic fibrosis," *New England Journal of Medicine* 331 (1994): 637–642.

115 *one research team has found a way:* The possibility of inserting DNA into mouse sperm is reported by Ralph Brinster in a forthcoming study. At the time of writing, there are also announcements of new strategies for harvesting human stem cells (from the research team of J. Scadden).

117 *Children with SCID:* For an early account of the course of the gene replacement therapy, see R. M. Blaese et al., "Treatment of severe combined immunodefi-

ciency disease (SCID) due to adenosine deaminase deficiency with CD34+ selected autologous blood cells transduced with a human ADA gene," *Human Gene Therapy* 4 (1993) 521–527. Michael Blaese has supervised the treatment of the two girls who were the original recipients of gene replacement therapy and has worked with Donald Kohn in treating the three boys. Because the boys could be diagnosed in advance, they were born into a protected environment, and cells from placental blood that are rich in stem cells, and which are usually discarded, could be removed, modified, and reinjected. I am extremely grateful to Dr. Blaese for a telephone conversation in which he provided me with as yet unpublished details of the course of the therapy.

However, despite the initial enthusiasm for the results with the SCID children, many biomedical scientists have recently urged caution. They point out that the children still receive shots of the enzyme standardly given to other patients with their form of SCID, and note the difficulties of deciding just how much of the children's improved functioning is to be traced to the gene replacement therapy. Blaese and Kohn hope to cut the enzyme injections and demonstrate the effectiveness of the gene replacement, but, at the time that this book went to press, the prevailing sentiment appeared to be one of retreating from proliferating clinical trials. (See *Science* 269 [1995]: 1050–1055).

118   *A French Canadian woman:* Details of this case are given in M. Grossman et al., "Successful ex vivo gene therapy directed to liver in a patient with familial hypercholesterolemia," *Nature Genetics* 6 (1994): 335–341.

120   *there is also a newer spray:* This is outlined in R. C. Boucher et al., "Gene therapy for cystic fibrosis using E1-deleted adenovirus: a trial in the nasal cavity," *Human Gene Therapy* 5 (1994): 615–639. Unfortunately, it now appears that there is a risk that the adenovirus vector triggers immune reactions, and, as with SCID, biomedical researchers have recently been adopting a more cautious assessment of gene replacement therapy for CF. (See *Science* 269 [1995]: 1050–1055, especially 1052–3).

120   *The semiofficial catechism:* See, for example, D. Suzuki and M. Knudtson *Genethics* (Cambridge, Mass.: Harvard, 1989), and N. Wivel and L. Walters, "Germ-Line Gene Modification and Disease Prevention: Some Medical and Ethical Perspectives," *Science* 262 (1993): 533–538.

122   *The questions of risk should be confronted directly:* Here I concur with the general approach taken by John Harris in *Wonderwoman and Superman: The Ethics of Human Biotechnology* (Oxford: Oxford University Press, 1992). However, Harris and I differ in our evaluations of the kinds of genetic interventions that are likely in the next decades, and many of his scenarios seem to me to rest on very optimistic ideas about possibilities of gene replacement.

CHAPTER 6   *The New Pariahs?*

128   *why we value the privacy of information about ourselves:* My own reflections on privacy have been influenced by the ideas of J. J. Thomson, "The Right to Privacy," *Philosophy and Public Affairs* 4 (1975): 295–314; T. M. Scanlon, "Thomson on Privacy," *Philosophy and Public Affairs* 4 (1975): 315–322; and J. Reiman, "Privacy, Intimacy, and Personhood," *Philosophy and Public Affairs* 5 (1976): 26–44.

130   *During the 1970s:* A lucid account of the policy of excluding those with sickle-cell trait from the Air Force Academy can be found in Troy Duster, *Backdoor to Eugenics* (New York: Routledge, 1990), pp. 24–28.

133   *Matters are different:* The arguments that follow are similar to lines of reasoning developed by Norman Daniels, "The Genome Project, Individual Differences, and Just Health Care," in *Justice and the Human Genome Project,* eds. Timothy Murphy and Marc Lappé (Berkeley: University of California Press, 1994), pp. 110–132.

136   *the right strategy is surely to face it directly:* Sometimes those who oppose mandated universal health care grant the argument in principle, but invoke economic or social constraints that effectively rule out the possibility of solving the problem. Already many Americans have testified eloquently to the difficulties they have faced when a genetic test brought bad news to them or to one of their relatives. At a recent meeting in Bethesda, Maryland, organized by the National Action Plan on Breast Cancer, several women related their experiences when they learned that they carried alleles that placed them at high risk for breast cancer: They were denied coverage for treatments (prophylactic mastectomies) that they, and their medical advisors, regarded as urgently needed; they lost their jobs; they lost their health coverage. I hope that the courage displayed by these women, and the many like them, will ultimately inspire a grassroots movement to demand that the principled solution be implemented in practice.

Plainly, those in the business of health insurance will resist this change, for it means the end of a profitable line of business. As with other major changes in the structure of society, there are bound to be economic disruptions. When swords are turned into plowshares, when slavery is abolished, and when divorce is made readily available, people who have previously been in certain occupations—soldiers, slave traders, private detectives—have to change their careers. There may be a good argument for helping them to new occupations. But there is no good reason to block social justice simply to sustain those who now profit from an unjust situation.

136   *an interesting, and realistic, thought experiment:* Philosophers will recognize the considerations that follow as connected with the social contract tradition in political philosophy, and, specifically, as recapitulating the Rawlsian idea of decisions behind a veil of ignorance. The classic source is John Rawls, *A Theory of Justice* (Cambridge, Mass.: Harvard, 1971). For another attempt to apply Rawls's seminal ideas to medical contexts, see Norman Daniels, *Just Health Care* (New York: Cambridge University Press, 1985).

142   *It is increased still further:* Use of genetic information to make decisions about hiring and firing is discussed by Dorothy Nelkin and Laurence Tancredi, *Dangerous Diagnostics,* 2d ed. (Chicago: University of Chicago Press, 1994), and by Ruth Hubbard and Elijah Wald, *Exploding the Gene Myth* (Boston: Beacon, 1993).

144   *At least one person working as an air traffic controller:* I am grateful to a genetic counselor for providing me with this information. To preserve patient confidentiality, we have agreed that the counselor should remain anonymous.

146   *Testing job applicants for workplace-specific risks:* In *Risky Business* (New York: Cambridge University Press, 1991), Elaine Draper argues that the practice is far from benign. She provides richly detailed accounts of many cases of workplace testing.

152   *Unfortunately, as is clear from recent discussions:* During the past decades, public imagination has been captured by many different kinds of claims about the biological determination of human behavior. For a variety of efforts to debunk such claims, see N. Block and G. Dworkin eds., *The IQ Controversy* (New York: Pantheon, 1976); Stephen Jay Gould, *The Mismeasure of Man* (New York: Norton, 1981); R. C. Lewontin, S. Rose, and L. Kamin, *Not in Our Genes* (New York: Pantheon, 1984); and P. Kitcher, *Vaulting Ambition: Sociobiology and the Quest for Human Nature* (Cambridge, Mass.: MIT Press, 1985). But perhaps the overwhelmingly negative response to the determinist theses of R. Herrnstein and C. Murray, *The Bell Curve* (New York: The Free Press, 1994) shows that the public mood in the United States has changed.

153   *a botched contrast to the successful Tay-Sachs program:* L. B. Andrews et al., *Assessing Genetic Risk* (Washington, D.C: National Academy Press, 1994), pp. 40–44, gives a measured discussion of the differences between sickle-cell and Tay-Sachs screening.

153   *Finally, a large chemical company:* Elaine Draper relates the details of the exclusionary policies of American Cyanimid in chap. 4 of *Risky Business.*

CHAPTER 7   *Studies in Scarlet*

157   *In 1983, in a small village:* Joseph Wambaugh provides an accessible and accurate account of the crimes and of the police inquiry in *The Blooding* (New York: Morrow, 1989).

158   *discovery that companies performing DNA analyses:* The most notable incident of problems in the reporting of DNA typing occurred in the case *New York* v. *Castro,* in which Lifecodes had great difficulty explaining the results in some lanes of the crucial gel. Eric Lander describes this case (in which he served as a witness) and discusses the general problems it poses, in "DNA Fingerprinting: Science, Law, and the Ultimate Identifier," in *The Code of Codes,* eds. D. Kevles and L. Hood (Cambridge, Mass.: Harvard, 1992), pp. 191–210. For the difficulty of deciding the sex of the person who had supplied the sample, see p. 200.

161   *which can serve as ideal "meaningless markers":* It is better for the markers to

be "meaningless," if possible, so that irrelevant further information cannot be extracted from DNA profiles. For the reasons given in the text, nonfunctional—"meaningless"—DNA will show most variation from person to person. So, happily, the best markers for purposes of identification are also the least invasive.

163   *belying the tactic of multiplying the probabilities:* This point was made very forcefully by R. C. Lewontin and D. Hartl in "Population Genetics in Forensic DNA Typing," *Science* 254 (1991): 1745–1750. It is also endorsed by authors much more sympathetic to forensic uses of DNA. See for example V. McKusick et al., eds., *DNA Technology in Forensic Science* (Washington, D.C: National Academy Press, 1992), especially pp. 78–82. The judgment of the National Academy report is that failures of independence can be circumvented in ways I shall consider later in the text.

164   *Inbreeding causes trouble:* The root of the controversy surrounding the statistical arguments used to estimate probabilities of DNA match is the extent to which natural populations are inbred or, by contrast, thoroughly mixed—often discussed as the issue of "population substructure." R. C. Lewontin and D. Hartl ("Population Genetics in Forensic DNA Typing") maintain that we do not know the extent of significant population substructure. Other researchers, for example, B. Devlin et al. ("No excess of homozygosity at loci used for DNA fingerprinting," *Science* 249 [1990]: 1416–1420) and R. Chakraborty and K. Kidd ("The Utility of DNA Typing in Forensic Work," *Science* 254 [1991]: 1735–1739) suggest that available statistics on various subgroups of the American population show them to be well mixed. It is not clear to me how the large-scale analyses employed by the latter authors could touch the possibility that major groups contain some much smaller populations that are significantly inbred. Hence I do not believe that the worries raised by Lewontin and Hartl have been completely resolved.

165   *Perhaps the problem can be made more tractable:* The discussion that follows introduces informally an approach that has recently become orthodox in statistical DNA forensics in the United States, the use of the so-called Ceiling Principle. That principle is formulated and motivated in V. McKusick et al., eds., *DNA Technology in Forensic Science,* pp. 82–95. Very recently, people who had originally seemed to take quite different attitudes to DNA typing have joined together to applaud the Ceiling Principle. See E. S. Lander and B. Budowle, "DNA Fingerprinting Dispute Laid to Rest," *Nature* 371 (1994): 735–738.

166   *Imagine that a new prosecutor continues the case:* As the imaginary scenario of this paragraph reveals, it is quite possible for the Ceiling Principle to give estimates of the probabilities that are injurious to the defendant. In their defense of the principle ("DNA Fingerprinting Dispute Laid to Rest"), Lander and Budowle are much more concerned to defend it against objections that it is too *conservative* (that is, it doesn't provide prosecutors with opportunities to wave around sufficiently stupefying probabilities). However, they do mention the possibility that the principle may err in the opposite direction (point [5] on p. 737), countering

this suggestion by proposing that component subpopulations can be taken to be well mixed. I am more worried about that possibility than are Lander and Budowle because I don't think we know the extent to which actual human populations may contain pockets of inbreeding. The available statistics reveal mixing at a gross level but are compatible with a significant amount of internal structure. To calculate probabilities on the basis of the Ceiling Principle thus seems to me to transfer our initial uncertainty to the question of how confident we can be about the absence of pockets of inbreeding, and in circumstances in which a defendant's fate hangs on a decision, it appears unwise to have a general policy of dismissing possibilities we *think* to be remote as if they were irrelevant. However, in particular instances, prosecutors *may* be able to assemble cogent statistical arguments for thinking that appeals to the possibility of amounts of inbreeding sufficient to undermine their figures are genuinely strained.

167   *Prosecutors proceed on the assumptions:* Detailed scrutiny of the statistical basis of the DNA arguments did not surface in the murder trial of O. J. Simpson. The prosecution offered quite dramatically low estimates that anyone but Simpson could have left certain samples of blood found at the scene of the crime. Even in high-quality newspapers there was little discussion of the underlying calculations—and the impression given to the public was of an arcane science whose minute details were baffling and dry. Both the press and the defense seemed much more concerned with the possibility of an elaborate police conspiracy.

    Does this mean that the trial avoided the major issues about DNA testing? In one sense, the answer is clearly "Yes," for the exact force of the DNA evidence against Simpson cannot be assessed without careful presentation of the questions I have raised in the text. However, there is a special feature of the case which sets complex statistical analysis on the sidelines. The prosecution claimed a match between Simpson's DNA and samples found at the scene of the crime, *and* matches between his victims' DNA and samples found on Simpson's property. From the defense's perspective, these matches have to be viewed as independent. Hence, to undercut the prosecution's arguments, it would be necessary to plead significant population substructure in *both* cases, enough to raise the separate probabilities that others might have matched the samples to the point at which their product becomes significant. This would be an extremely difficult and technical argument to make, and would be vulnerable to the charge that it envisions quite implausible coincidences. Thus it is hardly surprising that the defense focused on much more comprehensible points about laboratory errors and police conspiracy.

170   *in addition to the very real chance that jurors will misunderstand:* The difficulty even well-educated people have in dealing with probabilities is well known. Even though it follows from the laws of probability that the chance that someone will have two characteristics is no greater than the chance of having one of the characteristics alone, people frequently make judgments that violate this principle. See, for example, R. Nisbett and L. Ross, *Human Inference: Strategies and Shortcomings of Social Judgment* (Englewood Cliffs: Prentice Hall, 1980), especially

chap. 7. There are thus important issues concerning just what jurors should be told if their verdicts are likely to correspond to the balance of evidence. I do not pursue these issues here.

171 *it possible to recognize that the shifting of relative position has occurred:* For a lucid discussion of "bandshifting," see Eric Lander, "DNA Fingerprinting: Science, Law, and the Ultimate Identifier," pp. 202–204.

171 *Stephen Jay Gould has lucidly analyzed:* Gould reports his re-creation of the craniometers' measurements in *The Mismeasure of Man* (New York: Norton, 1981), chap. 2. Gould's own confession of error-induced-by-preconception is noted on p. 66.

172 *Any country could try to construct a national DNA database:* The American National Committee on DNA Technology in Forensic Science considered the possibility of a national DNA database, and rejected it (V. McKusick et al. eds., *DNA Technology in Forensic Science,* pp. 121–122) partly because of economic considerations, partly because of ethical and social reasons. The argument given in the text should be viewed as an attempt to respond to the latter concerns and to motivate a more careful economic analysis.

177 *Alex Jeffreys's hopeful vision:* Jeffreys announced his vision in two seminal papers: A. J. Jeffreys et al., "Individual-specific 'fingerprints' of human DNA," *Nature* 316 (1985): 75–79; and P. Gill, A. J. Jeffreys, and D. J. Werrett, "Forensic Applications of DNA 'fingerprints,' " *Nature* 318 (1985): 577–579.

179 *anecdotes of New York prostitutes:* Here I am indebted to a conversation with Dorothy Nelkin.

CHAPTER 8   *Inescapable Eugenics*

187 *Pregnant women can discover the sex:* For discussion of the uses of amniocentesis and selective abortion in Northern India, see D. Kumar, "Should One Be Free to Choose the Sex of One's Child?" in *Ethics, Reproduction and Genetic Control,* ed. Ruth Chadwick (London: Routledge, 1992), pp. 172–182.

188 *The history of eugenics:* There are now a number of excellent book-length treatments of the history of eugenics. For eugenics in the United States and Great Britain, Daniel J. Kevles's *In the Name of Eugenics* (New York: Knopf, 1985) is both comprehensive and lucid. For Nazi Germany, I have relied upon Robert Proctor, *Racial Hygiene: Medicine Under the Nazis* (Cambridge, Mass.: Harvard, 1988) and Benno Müller-Hill, *Murderous Science: Elimination by Scientific Selection of Jews, Gypsies, and Others, Germany 1933–1945* (Oxford: Oxford University Press, 1988).

193 *Practical eugenics is not a single thing:* Although most of the literature on eugenics does not differentiate the many forms of eugenic practice, a similar analysis to that offered here is given in an illuminating essay by Diane Paul, "Eugenic Anxieties, Social Realities, and Political Choices," *Social Research* 59 (1992): 663–683.

194 *Henry Goddard's efforts:* Stephen Jay Gould provides a compelling account of Goddard's zeal in testing both immigrants and inmates of institutions in *The Mismeasure of Man* (New York: Norton, 1981), pp. 158ff. See also Kevles, *In the Name of Eugenics,* chap. 7.

194 *Carrie Buck and her sister:* The story of the classification of Carrie Buck and her daughter, Vivian, is movingly told by Gould, "Carrie Buck's Daughter" (in *The Flamingo's Smile* [New York: Norton, 1985], pp. 306–318). Kevles's discussion (*In the Name of Eugenics,* 110–112) is also excellent.

195 *Doris lived outside the asylum:* For the poignant facts about Carrie Buck's sister, Doris Buck Figgins, see Gould, *The Mismeasure of Man,* pp. 335–336.

198 *There have already been women:* Barbara Katz Rothman provides an extensive and sensitive discussion of reactions to amniocentesis and test results in *The Tentative Pregnancy* (New York: Viking, 1986).

201 *Anglo-Saxons lamented "the passing of the great race":* One of the most influential eugenic documents of the 1920s was Madison Grant's *The Passing of the Great Race* (published in 1916), which emphasized the need to preserve the purity of the highest (Nordic) "human stock."

CHAPTER 9 *Delimiting Disease*

207 *From 1900 to 1960 many homosexuals:* The struggle to change the classification of homosexuality as a disease is well described by Ronald Bayer in *Homosexuality and American Psychiatry* (New York: Basic Books, 1981).

208 *Answers to these questions:* The constructivist approach to disease stems from Thomas Szasz' critique of the notion of mental illness ("The Myth of Mental Illness," *American Psychologist* 15 [1960]: 93–118). Contemporary objectivism is best represented by Christopher Boorse ("On the Distinction between Disease and Illness," *Philosophy and Public Affairs* 5 [1975]: 49–68; "Health as a Theoretical Concept," *Philosophy of Science* 44 [1977]: 542–573).

210 *Objectivists must search out more subtle factors:* These points, and many of these examples, are noted by Christopher Boorse ("Health as a Theoretical Concept").

211 *the design of the Creator has given way:* The idea that functions are to be understood in terms of the pressures of natural selection is developed by Larry Wright (*Teleological Explanation* [Berkeley: University of California Press, 1976]). The main alternative proposal, that functions should be understood in terms of contributions to the goals of a containing system, is articulated in Robert Cummins ("Functional Analysis," *Journal of Philosophy* 72 [1975] 741–760). In "Function and Design," *MidWest Studies in Philosophy* 18 (1994), I have tried to develop a position that is sensitive to the merits of both views, and this position underlies the approach taken in the text.

213 *the recent suggestion that some male homosexuals:* This hypothesis is presented in Simon LeVay, *The Sexual Brain* (Cambridge, Mass.: MIT Press, 1993).

213 *Nuclei of the smaller size appear to be dysfunctional:* If men with smaller nuclei

reproduce less, then there would be selection against the trait of having a smaller nucleus in a male body, unless the trait were correlated with some other characteristic that promoted spread of the pertinent genes. This is an area in which speculation is rife (see P. Kitcher, *Vaulting Ambition: Sociobiology and the Quest for Human Nature* [Cambridge, Mass.: MIT Press, 1985], pp. 243–252, for some diagnoses), and for the sake of argument in the text, I take the simple and straightforward view that homosexual men would reproduce less and that, in consequence, there would be selection pressure against the alleged underlying neurological causes.

214 *There are surely some breakdowns in functioning:* Here I have been influenced by the ideas of T. M. Scanlon's "Preference and Urgency," *Journal of Philosophy* 72 (1975): 655–669.

CHAPTER 10   *Playing God?*

221 *Men and women who have overcome hereditary disabilities:* Marsha Saxton speaks eloquently for the disabled in "Prenatal Screening and Discriminatory Attitudes About Disability" in *Embryos, Ethics, and Women's Rights,* eds. E. Hoffman Baruch et al. (New York: Harrington Park Press, 1988), pp. 217–224.

226 *One person's rights may be overridden:* The approach sketched here was originally developed in a seminal essay by Judith Jarvis Thomson ("A Defense of Abortion," in *The Rights and Wrongs of Abortion,* eds. M. Cohen et al. [Princeton: Princeton University Press, 1974]. A major contemporary version of the approach is Frances Myrna Kamm, *Creation and Abortion* (New York: Oxford University Press, 1992). My own perspective on the problem follows the contrary line charted by Ronald Dworkin (*Life's Dominion* [New York: Knopf, 1993]). However, the writings of Thomson, Kamm, and others with similar views clear up numerous confusions that surround the abortion controversy.

227 *Abortion cannot be assimilated to trimming nails:* Barbara Katz Rothman makes it very clear how wrenching the decision to abort fetuses with Down syndrome can be. See *The Tentative Pregnancy* (New York: Viking, 1986).

228 *prior to twenty-six weeks' gestation:* Ronald Dworkin gives this figure in *Life's Dominion,* p. 22, citing Michael Flower, "Neuromaturation of the Human Fetus" (*Journal of Medicine and Philosophy* 10 [1985]: 237–251; see especially 246–248). Flower, in turn, relies upon research into the neurology of pain, in particular, H. J. Ralston, "Synaptic organization of spinothalamic tract projections to the thalamus, with special reference to pain" in *Advances in Pain Research and Therapy,* vol. 6, eds. L. Kruger and J. C. Liebeskind (New York: Raven, 1984), pp. 183–195. The neuro-developmental issues are complicated, but to the best of my knowledge, the current consensus is that this is indeed a conservative estimate of the time at which the connections *currently taken to be crucial to sentience* form. I am grateful to Patricia Churchland and Adina Roskies for helpful discussion of the topic.

228 *Thanks to the philosopher:* See Ronald Dworkin, *Life's Dominion,* chaps. 1–3.
229 *"begun in earnest":* The phrase is Dworkin's. See *Life's Dominion,* pp. 97–101.
237 *Yet there is a last version of the worry:* The line of argument I consider here was presented to me, in conversation, by Ruth Hubbard.

CHAPTER 11   *Fascinating Genetalk*

239 *"All lawyers have it":* Peter Mayle, *Acquired Tastes* (New York: Bantam Books, 1986), p. 32.
242 *Genetic determinism has its intellectual roots:* Many contemporary writers believe that the roots of genetic determinism are not purely intellectual. See, for example, R. C. Lewontin, S. Rose, and L. Kamin, *Not in Our Genes* (New York: Pantheon, 1984).
243 *Our genetically determined characteristics:* For discussing the issues with which I am concerned in this book, where the focus is on how we might intervene to make human lives go better, characterizing genetic determinism in terms of the invariance of characteristics across environments in which people could flourish is entirely appropriate. However, there are other topics for which a different definition (for example, one that did not simply consider environments in which people can thrive) would be preferable—as when we consider whether or not certain types of knowledge or practical ability (such as the knowledge or ability that underlies language learning) are genetically determined. I am grateful to Ned Block for pointing this out to me.
245 *Classical and molecular geneticists:* The practice of tagging genes by taking them to be "for" the phenotypic traits they influence is omnipresent in the history of genetics, pervading virtually every research paper and textbook. The strategy for reconstructing that practice sketched in the text is worked out in far more detail in K. Sterelny and P. Kitcher, "The Return of the Gene," *Journal of Philosophy* 85 (1988): 335–358.
247 *For example, there might be alleles "for" dyslexia:* Similar conclusions have been drawn by Richard Dawkins in *The Extended Phenotype* (San Francisco: Freeman, 1982).
249 *In the wake of the discovery:* The discovery is announced in Y. Zhang et al., "Positional cloning of the mouse *obese* gene and its human homologue," *Nature* 372 (1994): 425–431. The authors are far more restrained than the popular press in offering conclusions about the causes of obesity in human populations, but a remark early in the article—". . . there is evidence that body weight is physiologically regulated"—coupled with the identification of a gene has, predictably, misled the incautious. The media response to the *ob* gene is a paradigm of the fascinations of genetalk.
253 *Molecular geneticists usually steer with ease:* Many human molecular geneticists are admirably lucid in recognizing the complications involved in talk of cancer as

a genetic disease. An exemplary discussion occurs in the opening pages of M-C. King et al., "Inherited Breast and Ovarian Cancer: What Are the Risks? What Are the Choices?" *Journal of the American Medical Association* 269 (1993): 1975–1980.

254 *Using animal models to illuminate:* I have tried to diagnose the problems of identifying "rape" in ducks, scorpion flies, and human beings in *Vaulting Ambition: Sociobiology and the Quest for Human Nature* (Cambridge, Mass.: MIT Press, 1985), pp. 184–190.

255 *a much-discussed study:* The original article is Dean Hamer et al., "A Linkage Between DNA Markers on the X Chromosome and Male Sexual Orientation," *Science* 261 (1993): 321–327. Hamer has since provided a detailed and accessible account of his research in a book coauthored with Peter Copeland, *The Science of Desire: The Search for the Gay Gene and the Biology of Behavior* (New York: Simon & Schuster, 1994). It should be emphasized that Hamer has shown an *association:* Not all the men in his study who carried the pertinent markers are homosexual, and not all the homosexual men whom he studied carried the markers. Recently there have been some critical discussions of Hamer's work, as other researchers have failed to replicate his findings; Hamer responds by suggesting that the research design of the new experiments is sufficiently different that the work "cannot be interpreted to either refute or confirm our findings." See "NIH's 'Gay Gene' Study Questioned," *Science* 268 (1995): 1841.

257 *When many genes affect:* The difficulties of probing conditions in which many loci are implicated are thoroughly explained in E. S. Lander and N. J. Schork, "Genetic Dissection of Complex Traits," *Science* 265 (1994): 2037–2048.

257 *Following the chronicle:* For over two decades, Victor McKusick has patiently compiled information on the genetic basis of various human conditions, publishing successive volumes entitled *On Mendelian Inheritance in Man* (Baltimore: Johns Hopkins, 1st ed. 1976, currently in the 9th ed.). Comparing the entries for the diseases that have been successfully mapped (Huntington's, cystic fibrosis, and the like) with various behavioral disorders, over several editions, is instructive.

258 *Among male monozygotic twins:* For detailed data on concordance rates, see Dean Hamer and Peter Copeland, *The Science of Desire,* chaps. 5 and 6.

260 *Nobody knows very much:* Both Dean Hamer (*The Science of Desire*) and Simon LeVay (*The Sexual Brain* [Cambridge, Mass.: MIT Press, 1993]) have been very cautious in acknowledging our current ignorance about the mechanisms underlying the development of sexual preference. Their writings make it abundantly clear that they do not subscribe to any simplistic determinist scenario.

262 *Recent experimental studies of fruit flies:* I draw the account that follows from recent work by the *Drosophila* behavior geneticist Tim Tully, "Genetic Dissection of Learning and Memory in *Drosophila melanogaster*" in *Neurobiology of Learning, Emotion, and Affect,* ed. J. Madden (New York: Raven Press, 1991), pp. 29–66, and "Gene Disruption of Learning and Memory: A Structure-Function Conundrum?" (*Seminars in Neuroscience,* 1994).

263 *Politicians are prepared to pay:* I am grateful to Troy Duster for information about the political pressures to investigate the genetic bases of violent behavior, especially during the Bush administration.

264 *inspired by announcements of a "genetic basis":* The most widely cited study is H. G. Brunner et al., "Abnormal Behavior Associated with a Point Mutation in the Structural Gene for Monoamine Oxidase A," *Science* 262 (1993): 578–580. For all the reasons given in the text, I believe that any conclusions drawn from this research must be highly tentative and that the most that has been shown is that some unfortunate men in a particular Dutch family have a lowered tolerance of environmental stresses, so that they are relatively easily provoked into violence.

268 *the poor man at his gate (or, perhaps, in the dungeon):* For a perspective that focuses on the social causes of a relationship between poverty and crime, see Jeffrey Reiman, *The Rich Get Richer and the Poor Get Prison,* 3d ed. (New York: Macmillan, 1990).

CHAPTER 12   *Self-Dissection*

274 *One of the most thoroughly understood developmental processes:* The mechanisms of lambda phages are described with enough background for the general reader and enough detail for the cognoscenti in Mark Ptashne, *A Genetic Switch* (Oxford: Blackwell, 1992).

275 *As the philosopher Daniel Dennett:* See Dennett's essay "Skinner Skinned" in his *Brainstorms* (Cambridge, Mass.: MIT Press, 1978), especially p. 65.

277 *The impulse to pit human freedom:* Many scientists and philosophers have argued that psychological determinism—the thesis that all our beliefs, desires, intentions, and actions are caused—is incompatible with human freedom. In *Beyond Freedom and Dignity* (New York: Knopf, 1971), B. F. Skinner argued that the incompatibility dooms hopes for human freedom. By contrast, in *Being and Nothingness,* tr. Hazel Barnes (New York: Philosophical Library, 1956), Jean-Paul Sartre insisted that because we are free, our psychological life cannot be determined. As will become plain in the text, I reject both types of incompatibilism.

279 *One of the most incisive thinkers:* David Hume's classic discussion of human freedom can be found in the chapter "Of Liberty and Necessity" in his *Inquiry Concerning Human Understanding* (reprinted by the Library of Liberal Arts, Indianapolis: Bobbs-Merrill, 1955).

281 *Contemporary philosophers, building on Hume:* The account I develop here has obvious kinship with the proposals of Harry Frankfurt and Gary Watson; both Frankfurt's "Freedom of the Will and the Concept of a Person" and Watson's "Free Agency" can be found in *Free Will,* ed. Gary Watson (Oxford: Oxford University Press, 1982). My version of the Humean approach has also been influenced by the writings of Daniel Dennett (*Elbow Room: Varieties of Free Will Worth Wanting* [Cambridge, Mass.: MIT Press, 1984]) and Susan Wolf (*Freedom Within Reason,* [New York: Oxford University Press, 1990]).

CHAPTER 13 *The Quality of Lives*

287 *Fundamental to the approach:* The perspective I try to articulate in this chapter has been influenced by the writings of a number of contemporary philosophers. In particular, I have drawn on the writings of John Rawls (*A Theory of Justice* [Cambridge, Mass.: Harvard, 1971]), Ronald Dworkin (*Life's Dominion* [New York: Knopf, 1993]), Norman Daniels (*Just Health Care* [New York: Cambridge, University Press, 1985]), James Griffin (*Well-Being* [Oxford: Oxford University Press, 1986]), and Dan Brock ("Quality of Life Measures in Health Care and Medical Ethics," in *Life and Death,* D. Brock [New York: Cambridge, 1993], pp. 268–324).

293 *the "Beethoven argument":* John Harris has offered an excellent discussion of a version of the argument offered by George Steiner and Germaine Greer on BBC television. See *Wonderwoman and Superman* (Oxford: Oxford University Press, 1992), pp. 179–182.

294 *We should be mindful of the woman:* For this example, see Ronald Dworkin, *Life's Dominion,* pp. 184, 210–211.

301 *If it is appropriate to determine the fate:* My discussion is influenced by Ronald Dworkin's sensitive and subtle connection of issues about "the edges of life," his linking of abortion and euthanasia, in *Life's Dominion.* I have also learned from Margaret Pabst Battin, *The Least Worst Death* (New York: Oxford University Press, 1994).

305 *The great preponderance:* Exactly how much money is spent in struggling against life's terminus has been a matter of debate for more than a decade. Paul Menzel notes that in 1975 60 percent of the money spent on health care in the United States "was spent on people who died within one year" (*Medical Costs, Moral Choices* [New Haven: Yale University Press, 1983], p. 186). Various people have responded to similar statistics by pointing out that large sums of money are often spent to try to avert death in cases where the outcome is quite uncertain (see, for example, Anne Scitovsky, "'The High Cost of Dying': What Do the Data Show?" *Milbank Quarterly* 62 [1984] 591–608): Doctors don't know, in advance of committing themselves to the expensive surgery, that the patient is going to die. However, if the suggestions of the text are correct, we should be asking questions not just about the chances of survival, but also about the expected quality of life that would follow expensive procedures. Ironically, in a recent response to those who believe American medicine could cut costs by modifying policies for treating elderly patients, "The Economics of Dying—The Illusion of Cost Savings at the End of Life" (*New England Journal of Medicine* 330 [1994]: 540–544, Drs. E. J. and L. I. Emanuel estimate the savings at "only" $29.7 billion; further, in replying to their critics, they concede that huge savings could be achieved by changing our attitudes toward the extension of the lives of the elderly. So I believe that the arguments of the text about the possibility of financing a large-scale program of preventive measures by cutting expenditure at the end of life are in accordance with the economic realities.

CHAPTER 14   *An Unequal Inheritance*

310   *These questions have been an important part of public debate*: The root of much opposition to contemporary molecular medicine seems to me to stem from a perceived contrast with social inaction. See, in particular, the writings of Ruth Hubbard and Richard Lewontin (Ruth Hubbard and Elijah Wald, *Exploding the Gene Myth* [Boston: Beacon, 1993]; R. C. Lewontin "The Dream of the Human Genome," in his *Biology and Ideology* [New York: HarperCollins, 1992], pp. 61–83). In the terms I use throughout this chapter, Hubbard and Lewontin are paradigm idealists; I suspect that virtually all the public defenders of the Human Genome Project and of the significance of molecular medicine are inclined to be pragmatists. However, since the issues have not previously been raised in the way I pose them here, it is impossible to do more than speculate about allegiances to pragmatism.

312   *What is the extent of our obligation:* The subsequent discussion is influenced by contemporary work in moral and political philosophy. In particular, I owe a considerable debt to the writings of John Rawls (*A Theory of Justice* [Cambridge, Mass.: Harvard, 1971]), T. M. Scanlon ("Preference and Urgency," *Journal of Philosophy* 72 [1975]: 655–669), and especially Thomas Nagel (*Mortal Questions* [Cambridge: Cambridge University Press, 1979]; *Equality and Partiality* [New York: Oxford University Press, 1991]).

317   *In a moving book:* Alex Kotlowitz, *There Are No Children Here* (New York: Doubleday, 1991). Similar points to those made in the text could also have been defended by focusing on an equally penetrating demonstration of how American society currently fails many of its children, Jonathan Kozol's *Savage Inequalities: Children in America's Schools* (New York: Crown, 1991).

325   *We are forced to confront the eugenic question:* Some central issues about population size and the quality of human lives are posed and explored with great clarity and insight in the fourth part of Derek Parfit's *Reasons and Persons* (Oxford: Oxford University Press, 1984).

Glossary

allele   The form of the genetic material at a locus. For example, many different alleles are known at the locus associated with cystic fibrosis; some pairs of alleles give rise to the disease, other pairs allow people to function normally.

amino acids   The units out of which proteins are built. There are twenty amino acids that are commonly found in the proteins of living organisms.

bases   The elementary units that occur in DNA molecules and that determine the specific forms of DNA molecules. There are four such bases: adenine (A), cytosine (C), guanine (G), and thymine (T). Bases are sometimes referred to as "nucleotides."

Canavan's disease   A degenerative disease that strikes infants, leading to decay of the nervous system and early death.

chromosome   A long threadlike structure within a cell that stains distinctly and can be observed with the light microscope. Genes are segments of chromosomes, although chromosomes typically contain long regions between genes. In most organisms the chromosomes are made of DNA. Chromosomes occur in pairs, and human cells (except for gametes) usually contain twenty-three pairs of chromosomes.

cloning   The technique of making multiple copies of a segment of DNA by inserting that segment into an organism (such as a bacterium or a virus) that will reproduce itself on a grand scale, thereby replicating the inserted DNA along with its own genetic material.

coding region   A segment of DNA that is transcribed and translated to produce a chain of amino acids that is part of a protein.

codon   A segment of DNA, three bases in length. Each codon (except for *stop codons*) is transcribed and translated to form a particular amino acid. The amino acid formed depends on the order of the bases in the codon.

cystic fibrosis   A genetic disease most common in people of Northern European extraction. Cystic fibrosis (or CF) patients used to die young from a variety of digestive and pancreatic ailments. At present, the most common source of problems results from the buildup of mucous secretions in the lungs, with consequent bacterial infections.

cytoplasm   The part of the cell lying outside the cell nucleus.

DNA   Deoxyribonucleic acid. DNA is the genetic material in most organisms (the exceptions are some viruses whose genetic material is RNA). DNA molecules consist of a pair of sugar-phosphate backbones that wind around one another (making a double helix), from which bases (or nucleotides) jut inward. The particular sequence of bases along a strand determines the individual character of the DNA molecule.

enzyme   A catalyst for a chemical reaction within a cell. Almost all known enzymes are proteins.

eukaryotes   Organisms whose cells have a nucleus within which the chromosomes are contained. "Higher" organisms, ranging from yeast to reptiles, fish, flies, and mammals, are eukaryotes.

exons   The parts of a gene that are represented by the corresponding amino acids in the finished protein. See also "introns."

Fragile X syndrome   A genetic disease in which the X chromosome has a tendency to break during mitotic divisions, so that the descendant cells miss part of the genetic material. Boys are affected more strongly than girls, and suffer moderate to severe mental retardation, as well as having behavioral problems.

gametes   The sex cells, sperm in males and ova in females. Gametes have only one member of each chromosome pair.

gel electrophoresis   A technique crucial to mapping and sequencing, in which DNA fragments are allowed to migrate through an agarose gel matrix under the influence of an electric current. Larger fragments move more slowly than smaller fragments, so that different pieces of DNA are separated in the gel.

gene   Classically, the unit of heredity. Before the advent of molecular methods, genes were typically identified through their influence on the observable characteristics of organisms. Genes are scattered along chromosomes and are usually segments of DNA (the exceptions are genes in some viruses, whose genetic material is RNA). Today genes are often conceived as DNA segments that code for proteins or, in some cases, for long chains of amino acids that are major parts of proteins.

gene replacement therapy   A strategy for attempting to ameliorate a genetic condition by inserting genetic material into a person's body. This is carried out in one of two ways: either by removing cells from the person's body, genetically modifying them, and then replacing them, or by injecting a specially constructed virus that will carry genetic material to the appropriate cells. At present, gene replacement therapy is in its infancy and has been tried only in a handful of cases.

genetic code   The scheme according to which particular three-base segments of DNA (codons) are associated with specific amino acids.

genetic marker   An identifiable piece of DNA that can be located in a definite chromosomal region. Genetic markers serve as signposts in the project of mapping. Prime examples of genetic markers are RFLPs and VNTRs.

genome   The totality of the genetic material of an organism. Molecular geneticists often think of a species-typical genome—as, for example, in the project of mapping the human genome. This usage envisages specifying the arrangement of loci along chromosomes that is typical of members of the species, a genetic parallel to the anatomical representation of the bodily structure of organisms of a species.

genotype   The combination of alleles present at a locus, or a number of loci. Context determines just which loci are pertinent: For example, in a discussion of cystic fibrosis, geneticists may identify a number of genotypes, distinguishing various combinations of alleles at the CF locus.

heterozygous   An organism with two different alleles at a locus is heterozygous at that locus. Heterozygous organisms are often called "heterozygotes."

homozygous   An organism with two copies of the same allele at a locus is homozygous at that locus. Homozygous organisms are often called "homozygotes."

Huntington's disease   A neurodegenerative disease that typically strikes people between thirty and fifty years of age. The genetic basis of the disease has been identified, but there are no known ways of preventing the degeneration and death.

Hurler syndrome   A genetic disease that causes disruption of cognitive development in early childhood and death, usually by age ten.

hypercholesterolemia (also cholesterolemia)   The condition of having elevated levels of LDL ("bad") cholesterol.

introns   The parts of a gene not represented by amino acids in the final protein. In eukaryotes, parts of the initial RNA transcript are excised before the mRNA leaves the cell nucleus. The pieces that are cut out correspond to the introns.

Lesch-Nyhan syndrome   A genetic disease affecting boys that produces both mental retardation and compulsive self-mutilation.

locus   A place on a chromosome (or chromosome pair) occupied by a gene (or by two alleles of the same gene).

map, genetic   A representation of the order of genes along a region of a chromosome, a whole chromosome, or several chromosomes.

map, physical   A collection of identifiable fragments of DNA from a chromosome region, a whole chromosome, or several chromosomes, together with a specification of how they are arranged along the chromosomal region (whole chromosome, chromosomes, etc.).

mapping   The enterprise of producing genetic and physical maps.

meiotic division   The cell division that produces gametes. Meiotic divisions yield cells with half the normal number of chromosomes.

mitotic division   The cell division that produces descendant bodily cells (somatic cells). In mitotic division, the normal number of chromosomes is preserved.

monozygotic twins   Twins who develop from the same fertilized egg (zygote) and therefore share all the genetic material in the cell nucleus. "Identical" twins are typically monozygotic.

mRNA   Messenger RNA. After transcription of DNA, an RNA is formed in the nucleus. In eukaryotes, this RNA transcript is modified through the excision of some segments (those corresponding to the introns) to yield a messenger RNA. The mRNA leaves the nucleus and is translated (to form a protein) in the cytoplasm.

mutation   The process through which the form of the genetic material is altered. The results of the alteration are often called "mutations" or "mutant alleles."

myotonic dystrophy   A genetic disease, affecting both males and females, in which the muscles are weakened. The effects are variable.

nucleotides   The units out of which DNA molecules are built—adenine (A), cytosine (C), guanine (G), and thymine (T). Often called "bases."

nucleus   A structure within the cells of many organisms (eukaryotes) that contains the chromosomes.

open reading frame   A string of bases that continues a long way without a stop codon, and hence a candidate for a region encoding a protein.

phenotype   The manifest characteristics of an organism, such as eye color, body build, and sexual orientation.

PKU   Phenylketonuria. A genetic disease that results from the inability to metabolize the amino acid phenylalanine. Children with PKU who are left untreated become severely retarded.

positional cloning   A strategy for hunting for genes implicated in disease (or, more generally, in any phenotypic trait) that uses genetic and physical maps to identify a chromosomal region within which the target gene is to be found, explores that region for candidate genes, and ultimately, if successfully executed, isolates, clones, and sequences the gene. Prominent examples of positional cloning are the identification of genes implicated in Huntington's disease, cystic fibrosis, and early-onset breast and ovarian cancer.

protein Proteins are important constituents of all living cells, playing vital roles in the chemical reactions that enable organisms to function. Proteins are composed of amino acids.

recombination The process in which chromosomes belonging to the same pair exchange genetic material.

regulatory regions Parts of DNA that affect the timing of the transcription of genes.

restriction enzymes Molecules that cut DNA. Each restriction enzyme is associated with a particular sequence that it recognizes and that determines just where it will cut.

RFLPs Restriction fragment length polymorphisms. RFLPs occur at places in the genome where variations affect the fragments produced by restriction enzymes. If two chromosomes carry different alleles at a RFLP locus (that is, they have DNA sequences there that affect whether or where a restriction enzyme will cut), then exposing them to the pertinent restriction enzyme will produce identifiably different collections of fragments. RFLP loci serve as signposts in the construction of genetic maps.

RNA Ribonucleic acid. RNA molecules are similar to DNA molecules except that they are usually single-stranded and contain uracil (U) in place of thymine (T). RNA plays important roles in the processes of protein formation.

Sanfilippo syndrome A genetic disease producing both severe mental retardation and extremely aggressive behavior.

SCID Severe combined immune deficiency. Children born with SCID lack a properly functioning immune system. The genetic basis of some types of this disorder is known, and some SCID children have received gene replacement therapy.

sequencing The process of producing the sequence of nucleotides (A's, C's, G's, and T's) along a strand of DNA. The result is presented as a long string of letters, each of which is "A," "C," "G," or "T."

sickle-cell disease A genetic disease most common in people of African descent. When two copies of a mutant allele are present, red blood cells contain abnormal hemoglobin, causing them to become rigid under conditions of low oxygen and to block blood flow. In many cases, these periodic crises lead to early death.

somatic cell A cell that is not a sex cell.

stop codon A three-base segment of DNA that signals the end of transcription.

Tay-Sachs disease A genetic disease most common in Jews of Ashkenazi descent. Infants with Tay-Sachs suffer neural degeneration during the first year and invariably die before the age of four.

thalassemia   Like sickle-cell disease, thalassemia is a genetic disease in which abnormal hemoglobin interferes with the functioning of red blood cells. It is most common among people from the countries that ring the Mediterranean.

transcription   The process in which a DNA segment gives rise to a corresponding piece of RNA. Transcription begins the process of protein formation.

translation   The process in which a chain of amino acids is formed corresponding to the sequence of bases along an mRNA molecule. Translation is a key step in protein formation.

VNTRs   Variable number tandem repeats. VNTRs occur at places in the genome where a particular sequence (for example, GAA) is repeated a number of times, and where the number is variable in the population. For example, at a particular place on chromosome 1, different people might have different numbers of repetitions of the sequence GAA (one person might have 11 repeats, another 15); of course, a person might have a different number of repeats at the same place on his or her two copies of chromosome 1 (10 on one copy, 23 on the other, say). Like RFLPs, VNTRs are useful signposts in the construction of genetic maps.

zygote   A fertilized egg.

Acknowledgments

In December 1992 I received a letter from Declan Murphy, then in charge of scholarly projects at the Library of Congress, asking me if I would be interested in serving as a Senior Fellow at the Library, with the task of evaluating the Human Genome Project and its impact. I had already been thinking about the Human Genome Project and the revolution in molecular genetics of which it is a part. Some years before, a postdoctoral fellow at Scripps Clinic in La Jolla, Sylvia Culp, had decided to pursue a second Ph.D. in philosophy and had written her dissertation under my supervision. We had both originally envisaged that dissertation as a timely exploration of the ethical and social implications of the explosion of knowledge in human molecular genetics, but, in the end, Sylvia's interests went in different directions, and the book we both agreed was needed remained unwritten. So Declan's letter touched a chord, and I wrote an enthusiastic response.

In the academic year 1993–4, with generous partial leave from the University of California at San Diego and the support of the Library of Congress, I was able to take a close look at the genetic revolution and to think about its social impact. I wrote a report for the Library of Congress, and led a number of discussions in Washington, D.C. However, it was quickly apparent that the issues raised by the genetic revolution need to be understood and discussed by the general public. So, with the encouragement of those who had supported my term as a Senior Fellow at the Library, I set out to write this book.

In the course of my research and writing, I have incurred a large number of debts, and it now gives me considerable pleasure to record the many unselfish people who have helped make the book much better than it would otherwise have been.

First, I would like to thank Declan Murphy and Prosser Gifford (his successor at the Office of Scholarly Programs) for their support, advice, and encouragement. Claudette Friedman assisted with great intelligence and thoroughness in the organization of my research for the Library of Congress. Dan Drell of the United States Department of Energy provided me with all kinds of valuable suggestions, and our conversations invariably helped me to focus my thoughts more precisely. I would also like to thank Richard Atkinson (then Chancellor of the University of California at San Diego) and Stanley Chodorow (former Dean of Arts and Humanities at UCSD) for making it possible for me to accept the invitation from the Library of Congress.

Having the opportunity to supervise Sylvia Culp provided me with a firm basis for my research, and I am most grateful to Sylvia for her part in teaching me molecular biology as I taught her philosophy. At an early stage of my research, I also benefited from the wise counsel of T. M. Scanlon and Dan Kevles. A discussion with Dan Kevles and Jim Woodward was particularly instructive.

During the fall of 1993, I was able to travel to a number of centers around the United States where human molecular genetics is currently advancing at a rapid pace. Although eminent scientists were sometimes bemused by the prospect of a philosopher impending upon them, they were invariably hospitable and helpful. I am most grateful to David Botstein, Mario Capecchi, Larry Deaver, David Galas, Leroy Hood, Eric Lander, Robert Moyzis, Mark Nefs, Marvin Notowicz, Maynard Olson, Jasper Rine, Gerald Rubin, Tim Tully, and Raymond White. Some people went to such great lengths to assist me that they deserve special thanks: Anthony Carrano and his colleagues at Livermore, Raymond Gesteland at the University of Utah, Walter Gilbert at Harvard, and Jan Witkowski of the Banbury Center at Cold Spring Harbor.

Many scholars who have thought about the implications of human molecular genetics were kind enough to explain their perspectives to me. I had valuable conversations with Peter Achinstein, George Annas, L. L. Cavalli-Sforza, Robert Cook-Deegan, Norman Daniels, Stephen Jay Gould, Stephen Hilgartner, Ruth Hubbard, Eric Juengst, Evelyn Fox Keller, Diane Paul, Philip Reilly, Kenneth Schaffner, Elizabeth Thompson, Nancy Wexler, and Michael Yesley. Dick Lewontin and Dorothy Nelkin were, as always, founts of knowledge, ideas, arguments, and questions. Both Troy Duster and Barbara Katz Rothman opened my eyes to new ways of looking at some of the problems I have been addressing.

Closer to home, I learned much from colleagues and students. Russell Doolittle, Glen Evans, Ted Friedmann, Michael Kaback, Michael Levine, and Christopher Wills provided information and advice. Michael Rothschild read drafts of the entire book with his usual acuity, and his detailed suggestions have greatly improved the clarity and the rigor of many passages. The graduate students in my seminar on the implications of human molecular genetics—Kenna Barrett, Linda Derksen, Joan Esnayra, Ilya Farber, Bruce Glymour, Mara Harrell, Josh Jorgensen, Brian Keeley, Alex Levine, Keri Monda, Cris Phillips, Michael Selgelid, and Scott Sterling—all helped me to refine my ideas. Numerous discussions with Bruce Glymour were particularly valuable, and I am indebted to him and to Josh Jorgensen for research assistance. Michelle Williamson, of UCSD's Office of Learning Resources, transformed my rough drafts of the illustrations into things of far greater clarity and beauty, and I appreciate her help.

Dr. Marilyn Jones of Children's Hospital enabled me to appreciate the devastating consequences that may flow from abnormal alleles, and I am most grateful to her for several conversations in which she gave me the perspective of an internationally renowned pediatric dysmorphologist. Dr. Nasra Ansar of Point Loma Mental Health Facility was also extremely helpful in showing me how hard it is to combat the environmental forces that threaten to wreck the lives of young children. Dr. Jerry Bordin read an entire draft of the book, offering me many insightful comments and suggestions, and I am most grateful to him for expressing so well the point of view of a practicing physician.

Several people have given me prepublication copies of their articles or have talked with me on the telephone about particular issues. Dr. Randi Hagerman took the time to discuss her work on Fragile X syndrome with me. In response to an inquiry, Dr. G. R. Sutherland sent me (via e-mail from Australia) his assessment of current statistics about Fragile X. Dr. Michael Blaese helped me understand the current state of gene replacement therapy for SCID. I am indebted to Rayna Rapp for sending me copies of forthcoming essays on the impact of genetic counseling on people from different cultural backgrounds. Diane Paul kindly shared with me some of her important research on the history of PKU testing. Finally, I would like to thank a genetic counselor, who should not be identified here, for some valuable information.

Numerous people have sent me comments on drafts of parts of this book and on related material. Sylvia Culp, Dorothy Nelkin, and Ken Schaffner gave extremely careful readings of early versions. Patricia Churchland, Daniel Dennett, John Dupré, Alex Levine, Richard Lewontin, Elliott Sober, Miriam Solomon, Kyle Stanford, and Frank Sulloway provided me with extensive written comments that have inspired major improvements. Briefer, but valuable, suggestions have come from Ned Block, Dan Drell, Horace Judson, Eric Lander, Alex Rosenberg, and Jan Witkowski.

I would like to thank my agent, John Brockman, for his assistance in presenting my project. I feel fortunate that he caught the attention of Alice Mayhew and that I have benefited from her superb editorial judgment. Roger Labrie gave the penultimate manuscript an extremely close reading, sniffing out vagueness, ambiguity, and unclarity, and his numerous pieces of advice have transformed some sections of the book, while improving all.

Finally, two people deserve very special acknowledgment. During the past few years, I have learned an enormous amount about contemporary biology from my friend and colleague, William Loomis. Bill spent many hours with me when I was in the early stages of research and, more recently, he has read and commented extensively on several versions of many chapters. For his intelligence, his generosity, his learning, and the pleasure of all our discussions, I am enormously grateful. Like Bill, Patricia Kitcher has been a marvelously constructive critic, and she also has read numerous drafts. I have benefited greatly from her philosophical acumen and have appreciated her many suggestions. She deserves more thanks than printed words can express.

With so much help, this book ought to be free from every flaw. Surely it is not. The fault, of course, is mine.

Solana Beach, California
May 1995

Index

Figures are indicated by page numbers in *italics*

A

A (adenine), 29, *30,* 31
abortion, 72
    of Down syndrome fetus, 198–99,
        227, 357*n*
    euthanasia vs., 300–301, 361*n*
    fetal rights debates and, 225–29,
        322
    moral concerns about, 203, 219,
        221–38, 283
    as murder, 223, 225
    natural (spontaneous), 26, 113, 224
    neural development and, 228, 357*n*
    objective criteria for, 204
    quality-of-life considerations for,
        217–19, 233–38, 286–300
    for reduction of human suffering,
        223–25, 227, 283, 284
    religious views on, 221, 222–25, 228,
        233
    reproductive rights and, 82–83, 84–
        85
    for sex selection, 187, 206, 216–17,
        218, 355*n*
    social pressures and, 199

value-of-human-lives assessments
        and, 229–34, 298–300
acute thyroid deficiency (cretinism), 107
addiction, 280
adenine (A), 29, *30,* 31
adoption, 72
African Americans, Air Force Academy
        positions denied to, 130–31, 153,
        351*n*
Africans, sickle-cell anemia among, 106,
        130, 153, 154, 367
aggressive tendencies, genetic basis for,
        152, 247, 250, 264–68
aging:
    of breast tissue, 74
    sperm count and, 157
agriculture, recombinant DNA in, 46
AIDS, 17, 83–84, 112, 222
Air Force Academy, sickle-cell carriers
        excluded from, 130, 153, 351*n*
air traffic controllers, 144
alleles, 363
allergies, 215
Alzheimer's disease, 257
American Cyanamid, 153, 352*n*

American National Committee on DNA Technology in Forensic Science, 355*n*

American Psychiatric Association, 207

amino acids:
codons for, 34, *35*
defined, 363
mutations and, 36–38, *37*
proteins composed of, 33, 363
sequencing of, 95

amniocentesis, 23, 25, 355*n*, 356*n*

Andrews, L. B., 347*n*, 352*n*

anemia, sickle-cell:
African descent and, 106, 130, 153, 154, 352*n*, 367
attacks of, 242
blood transfusions for, 109
employment discrimination and, 130–31, 153, 351*n*
environmental factors and, 242, 244
point mutation for, 36, 242
sickle-cell trait vs. 130–31
thalassemia vs., 85, 109, 368

Angastiniotis, M., 343*n*, 348*n*

Annas, G., 347*n*

Aristotle, 286

Ashkenazi Jews, Tay-Sachs disease among, 25, 28, 153, 154, 343*n*, 367

autism, 57

autoimmune disorders:
AIDS, 17, 83–84, 112, 222
molecular medical interventions in, 111–12, 117–18, 124
severe combined immune deficiency, 107, 117–18, 119, 121, 349*n*–50*n*, 367

AZT, 84

B

Bacon, Francis, 238

bacteriophages, replication of, 274–75, 360*n*

Barnes, Hazel, 360*n*

Baruch, E. Hoffman, 347*n*, 357*n*

bases (nucleotides):
in codons, 34, *35*
defined, 363
dideoxy, 55, *58, 59*
mutations of, 36–38, *37*
in open reading frames, 93
pairing of, 29–31, *30, 32*

Battin, Margaret Pabst, 361*n*

Bayer, Ronald, 356*n*

Beckwith, J., 344*n*

Beethoven family, reproductive history of, 293, 361*n*

behavior, genetic links to, 254–69
animal models for, 254, 262–63
antisocial actions and, 263–69
diagnosis and, 257
environmental influences and, 254–55, 257, 258–68
in homosexuality, 255–56, 258–62
long-term memory and, 262–63
in polygenic conditions, 257
urgent social problems and, 263–64

Berg, Paul, 344*n*

beta-thalassemia, 108–9, 154

bilateral mastectomy, 74

bipolar affective syndrome (manic de-pression), 257

birthrates:
global overpopulation problems and, 324–25
of mothers in poverty conditions, 321–323

Blaese, R. Michael, 349*n*–50*n*

Blake, William, 11

Block, Ned, 352*n*, 358*n*
blood-clotting factors, 46, 107
blood transfusions, 85, 108–9
bone marrow cells, gene modification
 techniques used on, 115
Boorse, Christopher, 356*n*
Boucher, R. C., 350*n*
brain:
 of chimpanzees vs. humans, 99
 gene replacement therapies and, 115–
  116, 120
breast cancer:
 early onset of, 257
 health insurance system and, 351*n*
 mammograms for, 74, 346*n*
 mastectomies and, 74, 351*n*
 polygenic links to, 257
 predictive value of genetic testing for,
  62–63, 74
 pregnancy age linked to, 212
 recent genetic research on, 53, 345*n*
 self-examination for, 74
Brinster, Ralph, 349*n*
Brock, Dan, 361*n*
Brunner, H. G., 360*n*
Buck, Carrie, 194–95, 240, 322, 356*n*
Buck, Doris, 194, 195, 356*n*
Buck, Vivian, 194, 195, 356*n*
Buck v. *Bell,* 194
Budowle, B., 353*n*–54*n*
Bunyan, John, 87
Bush, George, 360*n*

C

C (cytosine), 29, *30,* 31
Caenorhabditis elegans (nematode
 worm):
 genome research on, 89–90, 91, 93–
  94, 100–101, 348*n*

nervous system of, 263
Canavan's disease, 14, 79, 84, 107, 192,
 230, 285, 363
cancer:
 breast, 53, 62–63, 74, 212, 257, 345*n*,
  346*n*, 351*n*
 colon, 53, 70, 73–74, 345*n*
 environmental influences and, 61, 62,
  73, 253
 family histories of, 62
 genes associated with, 53, 62, 344*n*,
  345*n*
 as genetic disease, 253, 358*n*–59*n*
 genetic testing for, 62–63, 70, 73–74,
  346*n*, 351*n*
 lifestyle adjustments for prevention of,
  73
 ovarian, 62, 74, 345*n*, 346*n*
 workplace hazards and, 16–17
 yeast genome model and, 94–95
cardiovascular disease:
 arterial cholesterol buildup and,
  62
 genetic testing for, 62–63
 hereditary predispositions for, 17
 lifestyle adjustments for prevention of,
  73
causation, human freedom vs., 277–
 283
Ceiling Principle, 353*n*–54*n*
cells, division of, 31, *32,* 366
CF, *see* cystic fibrosis
Chadwick, Ruth, 355*n*
chain termination sequencing method,
 54–56
Chakraborty, R., 353*n*
Chicago, Ill., children in housing projects
 of, 317–18
childbirth, breast cancer development
 and, 212

children:
    genetic testing of, 23, 75, 84, 346*n*–347*n*
    natural death of, 223, 229–30
    sense of self in, 288–89
    in urban housing projects, 317–18
Children's Convalescent Hospital, 13–14, 15, 309
chimpanzees, genome research on, 99
cholesterolemia, 62–63, 118, 119, 350*n*, 365
    see also hypercholesterolemia
chorionic villi sampling, 25
Christianity, and deaths of children, 223
chromosomes:
    criminal, 69
    defined, 363
    extra copies of, 26–27
    loci on, 27–28
    sex, 26
    typical number of, 26
chromosome 21, extra copy of, 24, 27, 71, 227
    see also Down syndrome
Churchland, Patricia, 357*n*
cloning, 40, 46, 327, 330, 333–34, 335–36, 338, 339, 340, 341, 342
    defined, 363
    positional, 53, 255, 256–57, 366–67
coding regions, 161, 363
codons, 34, 363
Cohen, M., 357*n*
Collins, Francis, 343*n*
colon cancer, 53, 70, 73–74, 345*n*
colonoscopy, 73
Columbia University, 90
computer records, privacy regulations for, 128
Congress, U.S., Immigration Act of, 189

Connor, J. M., 344*n*
constructivism, 208–9, 210, 216, 356*n*
control lanes, 171
Cook-Deegan, Robert, 344*n*
Copeland, Peter, 359*n*
Copernicus, Nicolas, 271
copper, 110
Cowan, Ruth Schwartz, 347*n*
craniometers, 171, 355*n*
cretinism (acute thyroid deficiency), 107
Crick, Francis, 29, 31, 33, 44, 344*n*
criminals:
    DNA technology in prosecution of, 157–79
    environmental influences on, 264–68, 360*n*
    extra Y chromosomes among, 69
    popular beliefs about, 152
cultural achievement, human genetic capacity for, 99–100
Cummins, Robert, 356*n*
Cypriots, thalassemia among, 25, 85, 236, 322, 343*n*, 348*n*
cystic fibrosis (CF), 19, 165, 256
    ethnic origin and, 40, 154, 165, 364
    genetic testing for, 39, 53, 60, 71, 256, 294
    health insurance coverage and, 19, 135
    incidence of, 40
    life expectancy and, 71
    lungs and, 40, 111, 119–20, 349*n*, 364
    quality-of-life assessments for, 299–300
    treatment of, 107, 108, 111, 120, 349*n*, 350*n*
    varying severity of, 71
cytoplasm, 364
cytosine (C), 29, *30*, 31

D

Daniels, Norman, 351n, 361n

Darwin, Charles, 103, 191, 211, 212, 213, 214, 271

Davenport, Charles, 188–89, 190, 196, 240

Dawkins, Richard, 358n

death:
  euthanasia, 294, 301–4, 361n
  premature, 229–30

Defense, U.S. Department of, 130–31, 143–44

deletion mutations, 36, 37

Dennett, Daniel, 275, 360n

deoxyribonucleic acid, see DNA

determinism, genetic, 239–69
  antisocial behavior and, 264–69, 360n
  behavioral traits and, 247, 254–69, 358n–60n
  cholesterol levels and, 248–49
  criminality and, 265, 360n
  dyslexia and, 247–48, 358n
  environmental influences and, 241–68, 360n
  eugenic practices based on, 203
  genetic discrimination and, 129
  human freedom and, 277–83
  in Huntington's disease, 242–44, 252–53
  individualism and, 268–69
  molecular complexities lost in, 239–42, 245, 250
  obesity and, 249–50, 358n
  practical abilities and, 358n
  schizophrenia and, 250–51, 252, 257
  sexual preference and, 255–56, 258–62
  sickle-cell anemia and, 242, 244

in standard genetic and environmental backgrounds, 245–48

Devlin, B., 353n

diabetes, 46, 67

diagnosis:
  disambiguating, 109–10
  genetic testing utilized in, 26, 68, 344n

dialysis, 305, 306

Dickens, Charles, 266

Dictyostelium discoideum, genome research on, 92, 94

dideoxy bases, 55, 58, 59

disability activists:
  geneticists challenged by, 192
  prenatal testing feared by, 85, 200, 221–22, 347n, 357n

disability insurance, genetic information and, 140–41, 142

disambiguating diagnosis, 109–10

discriminatory practices, genetic basis for, 127–55
  in disability insurance, 140–41, 142
  in employment decisions, 142–50, 319
  in health insurance premiums, 131–39
  in life insurance coverage, 139, 140–42
  misinformation and, 130–31
  privacy concerns and, 127–29
  against sickle-cell carriers, 130–31
  and social stigmatization, 151–55

disease, criteria for, 207–19
  functional capacity and, 210–14, 356n
  objectivist vs. constructivist views on, 208–9, 210, 216, 356n
  quality-of-life assessments in, 214–19
  urgency factors in, 214

diseases, genetic:
  autoimmune disorders, 107, 111–12,
    117–18, 119, 121, 349n–50n, 367
  brain function and, 115–16
  Canavan's disease, 14, 79, 84, 107,
    192, 230, 285, 363
  cancers as, 53, 62, 253, 344n, 345n,
    358n–59n
  cystic fibrosis, 19, 39, 40, 53, 60, 71,
    107, 108, 111, 119–20, 135, 154,
    165, 256, 299–300, 349n, 350n,
    364
  development of tests for, 38–40, 56
  dietary therapies for, 66, 67, 75, 81,
    107, 108, 244, 345n
  disambiguating diagnosis and, 109–
    110
  Down syndrome, 24, 26–27, 29,
    71–72, 110, 198–99, 205, 227,
    288–89, 290, 357n
  economic inequality and, 198
  environmental influences vs. genetic
    factors in, 60–61; see also gene-
    environment interactions
  ethnic tendencies toward, 25, 28, 40,
    85, 106, 130–31, 153, 154, 165,
    236, 343n, 347n, 352n, 364, 367
  eugenic concerns and, 190, 192,
    198–201, 202–3
  Fragile X syndrome, 57–60, 69–70,
    115, 122, 135, 288, 344n,
    345n–46n, 364
  Gaucher's disease, 110–11, 349n
  gene replacement therapy and, 108,
    113–23
  and genetic susceptibilities to dis-
    eases, 252–53
  genetic testing vs. treatments for, 24
  health insurance coverage and, 135;
    see also health insurance

  Huntington's disease, 39, 40, 53–54,
    56–57, 65, 73, 76, 79, 107, 115,
    122, 144, 242–44, 252–53, 256,
    344n, 347n, 365
  Hurler syndrome, 190, 192, 214, 285,
    365
  hypercholesterolemia, 62–63, 118,
    119, 350n, 365
  Lesch-Nyhan syndrome, 82–83, 84,
    115, 116, 120, 121, 223, 225, 227,
    230, 288, 365
  myotonic dystrophy, 57, 122, 290, 366
  phenylketonuria, 65–68, 73, 75,
    81–84, 107, 108, 110, 215, 244,
    272, 345n, 366
  proteins in, 110
  quality-of-life assessments and, 285,
    288–289, 299–300
  Sanfilippo syndrome, 190, 209, 216,
    223, 290, 367
  sequencing disorders and, 56–60
  severe combined immune deficiency,
    107, 117–18, 119, 121, 349n–50n,
    367
  sickle-cell anemia, 36, 85, 106, 109,
    130–31, 153, 154, 242, 244, 351n,
    352n, 367, 368
  Tay-Sachs disease, 25, 28–29, 38, 39,
    60, 79, 153, 154, 203, 205, 226,
    243, 288, 343n, 352n, 367
  thalassemia, 25, 85, 108–9, 154, 236,
    343n, 348n, 368
  treatment efforts for, 105–26
  Wilson's disease, 110–11, 349n
DNA (deoxyribonucleic acid):
  coding regions of, 161, 363
  copying of, 31, 32
  defined, 364
  double-helix molecular structure of,
    29–31, 30, 364

evolutionary process manifested in, 94–95

extracellular accumulations of, 111

genetic markers on, 352n–53n, 359n, 365

intergenic areas of, 41–43, 42, 92

isolation of functional segments in, 93

mutations in, 36–38, 37

pace of research efforts on, 15

as prosecutorial evidence, see deoxyribonucleic acid analysis, forensic use of

protein production and, 33–34, 35

recombinant, 44–46, 45

regulatory regions of, 41–43, 367

viral switching of replication patterns for, 274–75

DNA (deoxyribonucleic acid) analysis, forensic use of, 19–20, 157–79

availability of samples for, 178–79

Ceiling Principle used in, 353n–54n

convictions overturned by, 162

deterrent value of, 158

fingerprints vs., 159, 170, 179

in Leicestershire murders case, 157–158, 167–68, 169, 176

match probabilities in, 162–69, 172, 353n–55n

national database proposed for, 172–77, 319, 355n

polymerase chain reaction used in, 159, 160

reliability of, 158, 170–72, 319, 352n

Simpson prosecution's use of, 354n

technology employed in, 159–62, 160, 170–72, 352n–53n

DNAase, 111, 120, 349n

Dolly (Lamb number 6LL3), 327, 329, 330–31, 334, 339, 340, 342

Down syndrome, 205

abortion of fetuses with, 198–99, 227, 357n

extra chromosome in, 24, 26–27, 29, 227

prenatal testing for, 71–72, 110, 198–199

quality-of-life assessments in, 288–89, 290

retardation levels in, 27

Draper, Elaine, 352n

Drosophila melanogaster (fruit flies):

genetic research on, 47–49, 90–91, 100, 348n

long-term memory studied in, 262–63, 359n

drug addiction, 280

Dürrenmatt, Friedrich, 22

Duster, Troy, 347n, 351n, 360n

dwarfism, 46, 107

Dworkin, G., 352n

Dworkin, Ronald, 228–29, 347n, 357n, 358n, 361n

dyslexia, 247–48, 358n

E

E. coli (Escherichia coli) bacteria, 274

genome research on, 46, 93, 94, 100

EcoRI, 45, 51

Einstein, Albert, 22

electroconvulsive therapy, 120

Elias, S., 347n

Emanuel, E. J., 361n

Emanuel, L. I., 361n

emotional experiences, molecular biology in, 272–77

employment, genetic information and, 142–55

for captive workforce, 16–17, 147–48

employment, genetic information and
(*cont.*)
insurance considerations and, 142
and qualifications for jobs, 144–45
workplace-specific risks and, 16–17,
143, 144, 146–50, *149,* 352*n*
England, DNA analysis used for murder
prosecution in, 157–58, 167–68,
169, 176
Enlightenment, 286
environment:
cancer predisposition and, 61, 62, 253
genetic influence combined with, *see*
gene-environment interactions
of poverty and neglect, 264–68,
309–10, 317–18, 360*n*
specific genotypes' risks and, 144,
146–50, *149*
standard, 107, 245–48
enzymes, 33, *35*
defined, 364
restriction, 44, *45,* 50, 52, 367
therapeutic compensation for, 110
Equal Opportunities Commission, U.S.,
143
erythrocytes (red blood cells), 106, 242
Escherichia coli (E. coli), 274
genome research on, 46, 93, 94, 100
ethnic groups, diseases associated with:
cystic fibrosis, 40, 154, 165, 364
sickle-cell disease, 106, 130, 153, 154,
352*n,* 367
social stigmatization increased by, 78,
347*n*
Tay-Sachs disease, 25, 28, 153, 154,
343*n,* 367
thalassemia, 25, 85, 236, 343*n,* 348*n*
ethnicity, eugenic policies and, 189, 190,
194, 201, 240, 355*n,* 356*n*

eugenics, 187–204
British social reformers' espousal of,
193, 201–2
compulsory genetic testing and, 82
contemporary fears of, 190–91,
192–93, 323
financial considerations in, 297–98,
321–23
four types of decisions involved in,
193, 197–98
gene replacement and, 125
genetic counseling as, 196
immigration policy linked to, 189,
240, 356*n*
initial development of, 191
laissez-faire, 195–201, 202–3
mental disabilities targeted by, 189,
194–95, 205, 240
naval families and, 188
Nazi programs of, 189–90, 193–94,
195, 240, 355*n*
prenatal testing and, 195–201, 283
as science distorted by cultural prejudice, 191–92, 240
scientific inaccuracies and, 194–95
selective abortion vs., 196–97
U.S. studies in, 188–89, 356*n*
eugenics, utopian, 201–3, 205–19
education required for, 202, 203, 217,
234
future scenario of, 205–7, 217
global population problems and,
324–25
quality-of-life considerations in,
214–19, 233–38, 284–97, 320–21
social inequality concerns in, 320–24
value-free criteria for diseases in,
207–14
value-of-life issues and, 298–300

Eugenics Record Office, 188, 194–95
eukaryotes, *42*, 89, 95, 364, 365
euthanasia, 294, 301–4, 361*n*
evolutionary process:
  genetic research techniques based on,
    94–95, 96, 97
  reproductive success promoted in,
    212–14
exons, *42*, 43, 364
experiments:
  knockout, 96–97
  preconceptions and bias in, 171, 355*n*

F

families, DNA analysis in reuniting of,
  157
fate map, 90
females:
  Fragile X syndrome in, 57, 364
  infanticide of, 187
  selective abortion of, 187, 206,
    216–17, 218, 355*n*
  sex chromosomes of, 26
  see also women
Ferguson-Smith, M. A., 344*n*
fetus, as person, 225–29, 232–33
Figgins, Doris Buck, 194, 195, 356*n*
fingerprints, 159, 170, 179
Flower, Michael, 357*n*
Forbes, M., 345*n*
forensics, genetic analysis in, *see* de-
    oxyribonucleic acid analysis, foren-
    sic use of
Fragile X syndrome:
  brain function and, 115
  gene replacement therapy for, 115, 122
  genetic testing for, 57, 69–70, 345*n*–
    346*n*

health insurance coverage and, 135
  in males vs. females, 57, 69, 364
  quality-of-life assessment for, 288
  symptoms of, 57–60, 70, 344*n*,
    345*n*–46*n*, 364
  trinucleotide repeats and, 122
frameshift mutations, *37*
Frankfurt, Harry, 360*n*
freedom:
  causation vs., 277–83, 360*n*
  quality of life linked to, 284
free market, health care system and,
  136
French Canadians, Tay-Sachs disease
    among, 28
Freud, Sigmund, 271, 272
Friedmann, T., 345*n*, 349*n*
frontal lobotomy, 120
fruit flies *(Drosophilia melanogaster):*
  genetic research on, 47–49, 90–91,
    100, 348*n*
  long-term memory studied in, 262–63,
    359*n*
Fuchs, H. J., 349*n*
function:
  disease and, 210–12, 213–14, 356*n*
  evolutionary view of, 212–13
functional traits, 212

G

G (guanine), 29, *30,* 31
Galton, Francis, 103, 191–92
gametes, 26
  in cell divisions, 31, 366
  defined, 364
gang rapes, 178
Gaucher's disease, 110–11, 349*n*
gel electrophoresis, 50, 52, 54, 364

gene-environment interactions, 28, 60–61, 241–69
  behavior and, 254–69
  cancer and, 61, 62, 73
  dyslexia and, 247–48
  Fragile X syndrome and, 346n
  high cholesterol levels and, 248–49
  Huntington's disease and, 242–44, 252–53
  identical twins studies of, 254–55
  in lambda phage replication switching, 274
  obesity and, 249–50, 358n
  phenylketonuria and, 244
  schizophrenia and, 250–51, 252, 257
  sexual preference and, 255, 258–62
  sickle-cell anemia and, 242, 244
  in standard genetic and environmental backgrounds, 245–48
  violent behavior and, 264–68, 360n
gene replacement, 113–26
  current techniques for, 114–19
  for cystic fibrosis, 120, 350n
  defined, 364–65
  for enhancement of natural traits, 123–126
  germline interventions vs. somatic replacements in, 120, 122–23
  for hypercholesterolemia, 118, 119, 350n
  for Lesch-Nyhan syndrome, 115, 116, 120, 121
  moral concerns about, 119–26
  nonhuman research models of, 96–99
  regulation of gene function in, 116–19
  for SCID, 117–18, 119, 121, 349n–50n
  social ramifications of, 124–25
  by stem cell modification, 115, 349n
  therapeutic uses of, 108, 113–23
  for trinucleotide repeat disorders, 122
  unforeseen consequences of, 120–23, 125
  viral vectors used in, 114–16
genes:
  defined, 364
  environmental influences and, see gene-environment interactions
  hunt for, 92–95
  protein formation and, 33–34, 35
genetic determinism, see determinism, genetic
genetic diseases, see diseases, genetic
genetic information:
  Air Force discrimination based on, 130–31
  disability insurance coverage and, 140–41, 142
  employment opportunities and, see employment, genetic information and
  health insurance coverage and, 131–139, 319
  life insurance coverage and, 139, 140–42
  privacy concerns about, 127–29
  social stigmatization based on, 151–155
genetic markers, 352n–53n, 359n, 365
genetic profiles, 159–77
  construction of, 161–62
  defined, 159
  of inbred populations, 164–67
  match probabilities calculated for, 162–69, 172
  national database proposed for, 172–177, 355n
  technological errors in, 170–72

genetic research:
  future projections on, 91–92, 100
  pace of, 15, 91–92, 100–101
  protein function and, 95–96
  see also genome, genome research
genetic testing, 65–86
  accessibility of, 77–78, 86, 198, 347*n*
  for cancer, 62–63, 70, 73–74, 346*n*,
    351*n*
  of children, 23, 75, 84, 346*n*–47*n*
  commercial interests in, 74
  counseling services for, 76–80, 319
  development of, 24–25, 40
  as diagnostic technique, 25, 68, 345*n*
  disadvantages of, 15–16
  disease treatment research vs., 24, 87–
    88
  for Down syndrome, 71–72, 110, 198–
    199
  emotional impact of, 65, 76–80, 347*n*
  employment affected by, *see* employ-
    ment, genetic information and
  ethnic groups and, 78, 347*n*
  false results in, 54, 66
  family genetic markers and, 53–54
  for Fragile X syndrome, 57, 69–70,
    345*n*–46*n*
  government prohibitions against, 143
  for heart disease, 62–63
  for Huntington's disease, 40, 53–54,
    56–57, 79, 256, 344*n*, 347*n*
  individual rights vs. public interests
    in, 80–86
  in vitro fertilization and, 113
  legally compulsory, 80–84
  lifestyle adjustments dictated by, 23,
    24, 73
  for myotonic dystrophy, 57
  of newborns, 23, 84

  for phenylketonuria, 65–68, 73, 75,
    81–82, 83, 84, 110
  predictive value of, 68–76
  prenatal, *see* prenatal testing
  privacy concerns about, 127–29, 173,
    175–76, 351*n*
  in schools, 75, 346*n*–47*n*
  social inequities and, 77–78, 79–80,
    86, 198, 319, 347*n*
  statistical support for, 68–72
  technology for development of, 56–63
  therapeutic research vs., 24, 87–88
genome, genome research:
  data banks on, 93–94
  defined, 365
  errors in, 101–2
  evolutionary process and, 94–95, 96,
    97
  on fruit flies, 90–91, 348*n*
  of *Haemophilus influenzae* Rd, 349*n*
  homologies as technique for, 94–95
  human uniqueness and, 99–100
  international collaborations on, 88, 90
  on nematode worm, 89–90, 91, 348*n*
  nonhuman models for, 89–91, 93–95,
    96, 103
  open reading frames for, 93, 94, 366
  pace of, 100–102
  technological advances in, 101
  on yeast, 89, 91, 348*n*
  see also Human Genome Project
genotype:
  defined, 28, 365
  social stigmatization based on, 151–55
germline, 120, 122–23
Gilbert, Walter, 88, 348*n*
Gill, P., 355*n*
glutamic acid, 242
Goddard, Henry, 189, 194, 196, 356*n*

Gould, Stephen Jay, 171, 352*n*, 355*n*, 356*n*

gout, 82, 116

Grant, Madison, 356*n*

Greek Orthodox Church, 85, 348*n*

Greer, Germaine, 361*n*

Griffin, James, 361*n*

Grossman, M., 350*n*

growth hormone, 46, 107

guanine (G), 29, *30*, 31

Guthrie, Woody, 39

Gypsies, Nazi racial policies against, 153, 190

H

Haemophilus influenzae Rd, genomic sequencing of, 349*n*

Hagerman, R., 344*n*

Haldane, J.B.S., 96

Hamer, Dean, 255, 257, 258, 359*n*

Harris, John, 350*n*, 361*n*

Hartl, D., 353*n*

Harvey, William, 210

Hawking, Stephen, 289, 293, 294

HDL (high-density lipoprotein) cholesterol, 118

health care system:
    access to, 78, 319, 351*n*
    allocation of expenditures in, 305–7, 361*n*

health insurance, 131–42
    breast cancer testing and, 351*n*
    cystic fibrosis as basis for denial of, 19
    equal access to, 135–36, 319, 351*n*
    free-market and, 136
    risk distribution and, 136–38
    social inequality and, 135–36, 140–42

    stigmatization from rejection by, 151–52
    two-tiered approaches to, 138–39, 140–41

heart, malformations of, 27

heart disease:
    arterial cholesterol buildup and, 62
    genetic testing for, 62–63
    hereditary predisposition for, 17
    lifestyle adjustments for prevention of, 73

hemoglobin, 85, 106, 109, 242, 367, 368

hemophilia, 46, 106–7

hepatitis, 46

Herrnstein, R., 352*n*

heterozygotes, 130, 365

high-density lipoprotein (HDL) cholesterol, 118

Hippocrates, 206, 207

Hitler, Adolf, 189, 190, 196

HIV, 17, 222

Holmes, Oliver Wendell, 194, 321

Holtzman, Neil A., 345*n*, 346*n*, 347*n*

homologies, 94–95

homosexuality:
    environmental influence vs. genetic factors in, 258–62
    genetic markers sought for, 152, 250, 255–56, 259, 359*n*
    hypothalamic nuclei and, 213, 356*n*–57*n*
    Nazi persecution of, 153, 190
    parental attitudes toward, 188
    as psychiatric disease, 207–8, 209, 356*n*
    selective abortion and, 206
    self-conception and, 272
    social stigmatization of, 151, 153, 188, 262

homozygotes, 130, 365

Hood, L., 344n, 347n, 348n, 352n

hormones:

cloning and, 46

human growth, 41, 107

hospice movement, 302

Housing and Urban Development, U.S.
Department of, 318

Hubbard, Ruth, 343n, 346n, 347n, 352n,
358n, 362n

human genome, number of genes esti-
mated for, 92

Human Genome Project:

critical assessments of, 88–89, 310,
311, 348n

government funding for, 21

international cooperation on, 88

official goals of, 88, 102, 343n

public discussion on impact of, 192,
362n

see also genome, genome research

human growth hormone, 46, 107

human lives, value of, 229–34, 298–300

Hume, David, 279, 281, 360n

Huntington's disease:

brain function and, 115

employment discrimination and, 144

environmental factors and, 107,
242–44, 252

family history of, 39

gene replacement therapy for, 115, 122

genetic determinism and, 242–44,
252–53

genetic testing for, 40, 53–54, 56–57,
79, 256, 344n, 347n

lack of treatment for, 73, 107, 365

progression for, 39, 365

prospective victim's knowledge of, 65,
73, 76, 347n

Hurler syndrome, 190, 192, 214, 285,
365

hypercholesterolemia, 62–63, 118, 119,
350n, 365

diverse causes of, 109

gene-environment interactions and,
248–49

gene replacement therapy for, 118

genetic testing for, 25, 79

hypertension, 209, 210

hypothalamus, 260

homosexuality and, 213, 356n–
357n

I

identical (monozygotic) twins:

concordance rate for homosexuality
in, 258–59

defined, 366

environmental vs. genetic factors studied in, 254, 258–59

identical DNA sequences of, 157, 173

Immigration Act (1924), 189

immigration policies, eugenics and, 189,
240, 356n

immune system disorders:

AIDS, 17, 83–84, 112, 222

molecular medical interventions in,
111–12, 117–18, 124

severe combined immune deficiency,
107, 117–18, 119, 121, 349n–50n,
367

inbreeding, genetic profile matches and,
164–65, 353n

India, sex selection practices in, 187,
191, 193, 203, 216–17, 218, 355n

individualism, genetic determination supported by, 268–69

individual reproductive freedom:
  eugenic education and, 202, 203
  of laissez-faire eugenics, 197, 198–201
  poverty birthrates and, 321–23
  prenatal testing and, 82–83, 84–85
  social consequences of, 200–201
inequality:
  in availability of genetic testing,
    77–78, 79–80, 86, 198, 319, 347n
  effectiveness of social interventions
    on, 314–16, 318
  of health-care access, 78, 319, 351n
  resource redistribution and, 313–15,
    316, 318, 320
  of social background, 309–12
  urban problems and, 263, 264–68,
    317–18
  utopian genetics faced with, 320–24
infanticide, female, in India, 187, 191,
  193, 203, 216–17, 218, 355n
infants:
  enzyme therapy for, 111
  genetic testing of, 23, 84
  smiles of, 274, 275–76
infectious diseases, immune system enhancement and, 124
insertion mutations, 36–38, 37
insulin, 46, 67
insurance, see specific types of insurance
intelligence:
  assessment of, 194–95
  cranial measurements and, 171
  Ellis Island immigrants screened for,
    189
  genetic markers sought for, 152, 188,
    250
  prenatal selection for, 188
intergenic DNA, 41–43, 42, 92
introns, 42, 43, 365

in vitro fertilization (IVF), 25, 113, 114,
  198
iron, excessive accumulations of, 109

J

Jeffreys, Alex, 157, 158, 177, 179, 355n
Jews:
  Nazi racial policies against, 153, 190,
    194, 227
  Tay-Sachs disease among, 25, 28, 153,
    154, 343n, 367
Johnston, M., 348n
Jones, Kenneth, 344n
Judson, Horace Freeland, 344n
jurors, DNA evidence presented to, 170,
  354n–55n
  see also deoxyribonucleic acid analy-
    sis, forensic use of

K

Kaback, M., 343n
Kamin, L., 352n, 358n
Kamm, Frances Myrna, 357n
Kevles, Daniel J., 344n, 347n, 348n,
  352n, 355n, 356n
Kidd, K., 353n
kidney dialysis, 305, 306
King, J., 345n
King, M-C., 346n, 359n
King, Patricia, 347n
knockout experiments, 96–97
Knudtson, M., 350n
Kohn, Donald, 350n
Kotlowitz, Alex, 317, 362n
Kozol, Jonathan, 362n
Kruger, L., 357n
Kumar, D., 355n

L

lambda phages, replication of, 274–75, 360n
Lander, Eric, 345n, 352n, 353n–54n, 355n, 359n
Lappé, Marc, 351n
Lawrence, Peter A., 348n
LDL (low-density lipoprotein) cholesterol, 62, 118, 248–49
left-handedness, 206
legal system, genetic evidence in, see deoxyribonucleic acid analysis, forensic use of
Leicestershire, England, DNA analysis used for murder prosecution in, 157–58, 167–68, 169, 176
Leicester University, 157
Lenz, W., 343n
lesbianism, 258
    see also homosexuality; sexual preference
Lesch-Nyhan syndrome, 230, 365
    brain function and, 115
    gene modification for, 115, 116, 120, 121
    hypothetical genetic testing for, 82–83, 84
    quality-of-life assessment for, 288
    self-mutilation in, 82, 120, 223, 225, 227, 365
leucine, 34
LeVay, Simon, 356n, 359n
Lewontin, R. C., 348n, 352n, 353n, 358n, 362n
Liebeskind, J. C., 357n
Lifecodes Corporation, 158, 172, 352n
life insurance coverage, 139, 140–42
lifestyle, modifications of, 23, 24, 73
liver function, 109

lives:
    quality of, see quality of life
    value of, 229–34, 298–300
living wills, 302
lobotomy, 120
loci, 27–28, 365
Loma Portal Mental Health, 309–10, 314
long-term memory, fruit flies study of, 262–63, 359n
low-density lipoprotein (LDL) cholesterol, 62, 118, 248–49
lung infections, DNAase therapy for prevention of, 111, 119–20

M

McKusick, Victor, 353n, 355n, 359n
Madden, J., 359n
malaria, 324
males:
    with extra Y chromosome, 69
    Fragile X syndrome in, 57, 89, 364
    hereditary naval predilection in, 188
    sex chromosomes of, 26
mallards, forced copulation of, 254
malpractice suits, 75
mammograms, 74, 346n
manic depression (bipolar affective syndrome), 257
mapping, 40, 43–54
    defined, 365–66
    of fruit-fly genome vs. human genome, 47–49
    genetic vs. physical, 56, 365–66
    by Human Genome Project, 88
    physical, 52, 56, 88, 366
    positional cloning techniques in, 52–53

mapping (*cont.*)
   recombinant techniques and, 44–49,
      48
   restriction fragment length polymor-
      phisms used in, 49–52, *51*
   sequencing techniques and, 54–56, *58,*
      59
mastectomies, prophylactic, 74, 351*n*
Mayle, Peter, 239, 245, 358*n*
medical care, economic restrictions on,
   78
meiotic divisions, 31, 366
memory, long-term, 262–63, 359*n*
Mendel, Gregor, 102, 191
menopause, 74
mental disorders:
   constructivist approach to, 356*n*
   eugenic programs directed at, 189,
      194–95, 205, 240
   homosexuality classified as, 207–8,
      209, 356*n*
mental hospitals, XYY population of, 69
mental retardation:
   in Down syndrome, 27
   in Fragile X syndrome, 57–60, 69–70,
      346*n,* 364
   in Lesch-Nyhan syndrome, 82
   in phenylketonuria, 66, 81, 108, 215,
      244, 366
Menzel, Paul, 361*n*
messenger RNA (mRNA), 34, *35, 42,* 43,
   366
mice:
   gene modification techniques tested
      on, 115, 119, 349*n*
   genome research on, 93, 94, 97, 101
   obesity studies on, 94, 249, 250, 358*n*
Miki, Y., 355*n*
Mill, John Stuart, 80

miscarriages, 26, 113, 224
mitotic divisions, 31, 366
Moen, Joan, 345*n*
molecular biology, molecular medicine:
   complexity of, 274–77
   current effectiveness of, 317
   development of, 15, 26, 29–31, 33–36
   disease criteria and, 207–19
   global perspective on, 324–25
   Human Genome Project and, 21,
      88–89, 102, 310, 311, 348*n*
   human self-concepts affected by,
      271–83
   social inequalities and, 309–25
monozygotic (identical) twins:
   concordance rate for homosexuality
      in, 258–59
   defined, 366
   environmental vs. genetic factors stud-
      ied in, 254, 258–59
   identical DNA sequences of, 157, 173
Morgan, Thomas Hunt, 47, 49, 90, 245
motherhood, social welfare programs
      and, 321–23
Motulsky, A., 343*n*
mRNA (messenger RNA), 34, *35, 42,* 43,
   366
Müller-Hill, Benno, 355*n*
multiple sclerosis, 25
Murphy, Timothy, 335*n*
Murray, Charles, 336*n*
muscular degeneration, quality-of-life
      issues and, 292–93, 294
muscular dystrophy, 289, 292
music, emotional responses to, 273–74,
   275, 276
mutations:
   defined, 28, 29, 366
   deletions, 36, *37*

inheritance of, 34–36
insertions, 36–38, *37*
miscarriages caused by, 113
point, 36, *37*
for Tay-Sachs disease, 28–29
three types of, 36, *37*
myotonic dystrophy, 57, 122, 290, 366

N

Nagel, Thomas, 362*n*
National Action Plan on Breast Cancer, 351*n*
National Football League, 130
National Health Service, British, 305
natural selection, 97
    for function, 210–11, 356*n*
    human cultural values vs., 213–14
Nazis:
    eugenics of, 153, 189–90, 193–94, 195, 240, 355*n*
    Jews persecuted by, 153, 190, 194, 227
Nelkin, Dorothy, 343*n*, 346*n*, 352*n*, 355*n*
nematode worm *(Caenorhabditis elegans):*
    genome research on, 89–90, 91, 93–94, 100–101, 348*n*
    nervous system of, 263
Netherlands, physician-assisted death in, 302
neurofibromatosis, 14, 214, 288
newborns, genetic testing of, 23, 84
New York v. *Castro,* 352*n*
Nichols, E., 345*n*, 349*n*
Nisbett, R., 354*n*–55*n*
Nogee, L. M., 344*n*
nomadism, genetic hypothesis on, 240

Northern India, sexist cultural values in, 187, 191, 193, 203, 216–17, 218, 355*n*
nucleotides (bases):
    in codons, 34, *35*
    defined, 363
    dideoxy, 55, *58, 59*
    mutations of, 36–38, *37*
    in open reading frames, 93
    pairing of, 29–31, *30, 32*
nucleus, 366
Nüsslein-Volhard, Christiane, 90
Nyhan, W. L., 345*n*

O

obesity:
    environmental influence vs. genetic factors in, 249–50
    genes associated with, 94, 188, 206, 358*n*
    societal attitudes toward, 208, 209
objectivism, 208–10, 356*n*
    functional capacity and, 210–14, 216
olfactory loci, 27–28
oncogene, 62
open reading frames, 93, 94, 366
organ transplants, 112
Orwell, George, 176
ovarian cancer, 62, 74, 345*n*, 346*n*
overpopulation, 324–25, 362*n*

P

pain:
    neurology of, 228, 357*n*
    pleasure vs., 289, 294–96
Parfit, Derek, 362*n*

Pascal, Blaise, 271

paternity, DNA analysis for establishment of, 157

Paul, Diane, 66, 67, 345n, 355n

PCR (polymerase chain reaction), 159, 160

penicillamine, 110

phages, replication of, 274–75, 360n

phenotype, 97
  defined, 33, 366
  genetic influence vs. environmental impact on, 28, 240–42

phenylalanine, 66, 108, 244, 272, 366

phenylketonuria (PKU):
  dietary adjustments for, 66, 67, 75, 81, 107, 108, 244, 345n
  effectiveness of treatment for, 66, 317
  genetic testing for, 65–68, 73, 75, 81–82, 83, 84, 110
  mental retardation in, 66, 81, 108, 215, 244, 366
  molecular deficiency at root of, 66, 272, 345n

physical maps, 52, 56

Pitchfork, Colin, 158, 167–68, 169, 176

PKU, see phenylketonuria

pleiotropy, 265

point mutations, 36, 37

polymerase chain reaction (PCR), 159, 160

Pope, Alexander, 225

population growth, 324–25, 362n

positional cloning, 53, 255, 256–57, 366–67

pregnancy:
  genetic testing during, see prenatal testing
  PKU special diet and, 66, 67
  workplace toxins and, 153

prenatal testing:
  abortion decisions and, 72, 82–83, 84–85, 221–22, 283; see also abortion
  for AIDS, 83–84
  for cystic fibrosis, 71
  disability activists' opposition to, 85, 200, 221–22, 347n, 357n
  for Down syndrome, 71–72, 110, 198–99
  equal access to, 15–16, 198, 199–200, 202
  for extra Y chromosome, 69
  for Fragile X syndrome, 69–70
  individual reproductive freedom and, 197, 198–200
  laissez-faire eugenics and, 196–200
  medical techniques for, 25
  predictive value of, 69–72
  refusal of, 80–84
  for sex selection, 187–88, 206, 207, 216–17
  testing of newborns vs., 84

prison populations, XYY males in, 69

privacy issues:
  genetic test results and, 127–29, 351n
  and national DNA database, 173–74, 175, 355n

Proctor, Robert, 355n

proteins:
  defined, 367
  gene modification therapy and, 114, 117–19
  therapeutic compensation for absence of, 110, 111, 114, 117–19
  X-ray crystallography used for research on, 95–96

protein synthesis:
  molecular process of, 33–34, 35

mutations and, 36–38, *37*
psychoanalysis, human self-concepts affected by, 271
psychological assessments, molecular explanations in, 274–77, 360*n*
Ptashne, Mark, 360*n*

Q

quality of life, 284–307
  abortion decisions and, 217–19, 233–38, 296–300
  for families of disabled children, 297–98
  freedom as criterion for, 284
  in fulfillment of central desires, 289, 291–95
  historical changes in views of, 286
  medical-resource allocations and, 298, 306–7
  pleasure vs. pain in, 289, 294–96
  pluralistic attitude toward, 286–87
  right-to-die concerns related to, 294, 301–4
  sense of self in, 287–90
  social inequality and, 320–21
  in specific disease conditions, 214–16, 285, 288–89
  three dimensions for assessments of, 289–97
  value-formation capacity as criterion for, 289–91

R

rabbits, environmental adaptations of, 211
radioactivity, atomic decay in, 277–78
Ralston, H. J., 357*n*

rape, 178–79, 254, 359*n*
Rapp, Rayna, 347*n*
Rawls, John, 351*n*, 361*n*, 362*n*
recessive alleles, 29
recombination:
  defined, 47, 367
  techniques of, 44–49, *48*
red blood cells (erythrocytes), 106, 242
regulatory regions, 41–43, 367
Reiman, Jeffrey, 351*n*, 360*n*
religious belief:
  against abortion, 221, 222–25, 228, 233
  Darwinian theory and, 271
  genetic testing and, 81, 83
  of good life, 286
  on physiological function, 210–11
  pluralistic tolerance toward, 286
  sexism cloaked in, 187
renal dialysis, 305, 306
reproductive freedom:
  eugenic education and, 202, 203
  of laissez-faire eugenics, 197, 198–201
  poverty birthrates and, 321–23
  prenatal testing and, 82–83
  social consequences of, 200–201
respiratory disease, mutation as cause of, 36–38
restriction enzymes, 44, *45*, 50, 52, 367
restriction fragment length polymorphisms (RFLPs), 50–52, *51*, 367
restriction sites, 50–52
retardation:
  in Down syndrome, 27
  in Fragile X syndrome, 57–60, 69–70, 346*n*, 364
  in Lesch-Nyhan syndrome, 82

retardation (*cont.*)
   in phenylketonuria, 66, 81, 108, 215,
      244, 366
retroviruses, 115–16, 121
RFLPs (restriction fragment length poly-
     morphisms), 50–52, *51*, 367
ribonucleic acid (RNA), 34, *35, 42*, 43,
     366, 367
Romanticism, 286
Rose, S., 352*n*, 358*n*
Roskies, Adina, 357*n*
Ross, L., 354*n*–55*n*
Rothman, Barbara Katz, 347*n*, 356*n*,
     357*n*
Rousseau, F., 346*n*

S

Saccharomyces cerevisiae (yeast),
     genome research on, 89, 91, 93–94,
     95, 100, 101, 348*n*
Sanfilippo syndrome, 190, 209, 216, 223,
     290, 367
Sartre, Jean-Paul, 360*n*
Saxton, Marsha, 347*n*, 357*n*
Scadden, J., 349*n*
Scanlon, T. M., 351*n*, 357*n*, 362*n*
scanning tunnel electron microscopy,
     101
schizophrenia, 250, 251, 252, 257
Schneider, J. A., 349*n*
schools, genetic testing in, 75, 346*n*–47*n*
Schork, N. J., 359*n*
Schubert, Franz, 273–74
SCID (severe combined immune defi-
     ciency), 107, 117–18, 119, 121,
     349*n*–50*n*, 367
Scitovsky, Anne, 361*n*
scorpion flies, forced copulation by, 254

self-conceptions, molecular biological in-
     fluence on, 271–82
   emotional responses and, 272–77
   human freedom and, 277–83, 360*n*
   as quality-of-life factor, 287–90
   sexual orientation and, 272
self-mutilation, 82, 120
sequencing, 40
   data banks on, 93–94
   defined, 367
   error rates in, 101–2
   evolutionary process seen in, 94–95
   genetic disorders associated with,
     57–60
   of *Haemophilus influenzae* genome,
     349*n*
   homologies in, 94–95
   Human Genome Project work on, 21,
     88–89, 102, 192, 310, 311, 348*n*
   intergenic areas vs. gene-rich regions
     in, 92–95
   for nematode genome, 90, 348*n*
   pace of research on, 100–101
   techniques for, 54–56, *58, 59*,
     348*n*–49*n*
   technological advances in, 101
   for yeast genome, 89, 348*n*
severe combined immune deficiency
     (SCID), 107, 117–18, 119, 121,
     349*n*–50*n*, 367
sex chromosomes, 26
sex differences, intelligence and, 152
sex selection, 187–88, 206, 207, 216–18
sexual preference:
   gene-environment interactions and,
     258–62
   genetic basis for, 152, 213, 250,
     255–56, 259, 359*n*
   self-conception and, 272

social stigmatization of, 151, 152, 153, 188, 262
see also homosexuality
Shaw, George Bernard, 193, 201, 202
sickle-cell disease:
African descent and, 106, 130, 153, 154, 352*n*, 367
attacks of, 242
blood transfusions for, 109
employment discrimination and, 130–131, 351*n*
environmental factors in, 242, 244
point mutation in, 36, 242
sickle-cell trait vs., 130–31
thalassemia vs., 85, 109, 368
Silverman, A., 344*n*
Simpson, O. J., 354*n*
Singer, Maxine, 344*n*
Skinner, B. F., 360*n*
skull measurements, 171
smallpox, 221
smell, sense of, 27
social stigmatization, genotype as basis for, 151–55
somatic cells:
defined, 26, 367
gene replacement therapy on, 120, 122
mitotic division for production of, 31, 366
soul:
in abortion considerations, 223–24, 228, 233
good life for, 286
sperm, gene modification technology for, 115, 349*n*
sperm count, age related to, 157
spina bifida, 285, 292
standard environments, 107, 245–48

Stanford-Binet intelligence test, 194
Steinberg, M., 349*n*
Steiner, George, 361*n*
stem cells, 115, 349*n*
Sterelny, K., 358*n*
sterilization, compulsory, 189, 194–95, 240
stigmatization, genotype as basis for, 151–55
stop codons, 34, *35, 37,* 93, 367
strokes, 62, 303
Sturtevant, Arthur, 47, 48, 49
Surfactant B, 36–38, 39, 344*n*
Sutherland, G., 346*n*
Suzuki, D., 350*n*
Szasz, Thomas, 356*n*

T

T (thymine), 29, *30,* 31, 34
Tancredi, Laurence, 343*n*, 346*n*, 352*n*
Tay-Sachs disease:
abnormal alleles paired in, 28–29
insertion mutation in, 38
in Jewish populations, 25, 28, 153, 154, 343*n*, 367
mutation varieties for, 60
progress of, 28, 226, 243, 288, 367
quality-of-life assessments for, 288
recessive allele in, 29, 39
sickle-cell screening and, 153, 352*n*
tests for carriers of, 38, 79, 203, 352*n*
T cells, 117–18
terminal disease:
euthanasia and, 301–4
health care expenditures for, 305–7, 361*n*
thalassemia:
beta-, 108–9, 154

thalassemia (*cont.*)
blood transfusions for, 85, 108–9
defined, 368
ethnic groups affected by, 25, 154,
343*n*, 368
Greek Orthodox Church program for,
85, 348*n*
reduction in incidence of, 85, 236, 322
sickle-cell disease vs., 85, 109, 368
thalassophilia, 188, 263
Thomson, Judith Jarvis, 351*n*, 357*n*
thymine (T), 29, *30*, 31, 34
tomatoes, genetic alteration of, 46
tooth decay, 210
transcription, DNA, 34, *35*, 368
regulation of, 41–43
translation, DNA, 34, *35*, 43, 368
trinucleotide repeats, genetic diseases
marked by, 56–57
Tully, Tim, 359*n*
tumor-suppressor genes, mutant, 205,
206
twins, identical (monozygotic):
concordance rate for homosexuality
in, 258–59
defined, 366
environmental vs. genetic factors
studied in, 254, 258–59
identical DNA sequences of, 157, 173
tyrosine, 66, 108, 244

U
uracil (U), 34
urban social problems:
crime and violence, 264–68, 360*n*
public resource commitment for,
317–18
utopian eugenics, *see* eugenics, utopian

V
vaccine production, 46
valine, 34, 242
variable number tandem repeats
(VNTRs), 52, 368
Venter, Craig, 348*n*–49*n*
Verdi, Giuseppe, 273–74, 276
violent behavior:
environmental influence vs. genetic
source for, 264–68, 360*n*
genotype sought for, 152, 360*n*
viral vectors, 114–16
Virginia Colony for Epileptics and Feebleminded, 194
viruses:
replication of, 17, 274–75
as vectors for gene modification,
114–16
VNTRs (variable number tandem repeats), 52, 368

W
Wald, Elijah, 343*n*, 346*n*, 347*n*, 352*n*,
362*n*
Walters, L., 350*n*
Wambaugh, Joseph, 352*n*
Watson, Gary, 360*n*
Watson, James D., 29, 31, 33, 44, 344*n*
Webb, Beatrice, 201
Webb, Sidney, 193, 201
Werrett, D. J., 355*n*
Wertz, Dorothy, 347*n*
Wexler, Nancy, 39–40, 344*n*, 347*n*
Williams, R. C., Jr., 344*n*, 349*n*
Wilmut, Ian, 329, 330, 331, 340
Wills, Christopher, 344*n*, 349*n*
Wilson's disease, 110–11, 349*n*
Witkowski, J., 346*n*